Tucholsky   Wagner   Zola   Scott   Sydow   Freud   Schlegel
        Turgenev   Fonatne
              Wallace   Fouqué
       Twain   Walther von der Vogelweide   Friedrich II. von Preußen
                     Weber   Freiligrath
                                Kant   Ernst   Frey
Fechner   Fichte Weiße Rose von Fallersleben   Richthofen   Frommel
        Engels   Fielding   Hölderlin
   Fehrs   Faber   Eichendorff   Tacitus   Dumas
                      Flaubert   Eliasberg   Ebner Eschenbach
        Maximilian I. von Habsburg   Fock   Zweig
 Feuerbach   Ewald   Eliot   Vergil
      Goethe   London
                      Elisabeth von Österreich
Mendelssohn Balzac   Shakespeare   Dostojewski   Ganghofer
                 Lichtenberg   Rathenau
       Trackl   Stevenson   Hambruch   Doyle   Gjellerup
Mommsen   Tolstoi   Lenz   Droste-Hülshoff
          Thoma   von Arnim   Hanrieder
 Dach   Verne   Hägele   Hauff   Humboldt
      Karrillon Reuter   Rousseau   Hagen   Hauptmann   Gautier
             Garschin   Baudelaire
            Defoe   Hebbel
    Damaschke   Hegel   Kussmaul   Herder
                     Descartes   Schopenhauer
Wolfram von Eschenbach   Darwin   Dickens   Rilke   George
   Bronner   Melville   Grimm Jerome   Bebel   Proust
      Campe   Horváth   Aristoteles
Bismarck   Vigny   Barlach   Voltaire   Federer   Herodot
                Gengenbach   Heine
  Storm   Casanova   Tersteegen   Gilm   Grillparzer   Georgy
               Chamberlain   Langbein   Gryphius
  Brentano   Claudius   Schiller   Lafontaine
  Strachwitz   Schilling   Kralik   Iffland   Sokrates
       Katharina II. von Rußland   Bellamy
                         Gerstäcker   Raabe   Gibbon   Tschechow
 Löns   Hesse   Hoffmann   Gogol   Wilde   Gleim   Vulpius
    Luther   Heym   Hofmannsthal   Morgenstern
      Roth   Heyse   Klopstock   Klee   Hölty   Kleist   Goedicke
 Luxemburg   Puschkin   Homer   Mörike
       Machiavelli   La Roche   Horaz   Musil
Navarra Aurel   Musset   Kierkegaard   Kraft   Kraus
  Nestroy Marie de France   Lamprecht   Kind   Kirchhoff Hugo   Moltke
      Nietzsche   Nansen   Laotse   Ipsen   Liebknecht
                Marx   Ringelnatz
   von Ossietzky   May   Lassalle   Gorki   Klett   Leibniz
                     vom Stein   Lawrence   Irving
   Petalozzi
        Platon   Pückler   Michelangelo   Knigge   Kafka
      Sachs   Poe   Liebermann   Kock
          de Sade Praetorius   Mistral   Zetkin   Korolenko

The publishing house tredition has created the series **TREDITION CLASSICS**. It contains classical literature works from over two thousand years. Most of these titles have been out of print and off the bookstore shelves for decades.

The book series is intended to preserve the cultural legacy and to promote the timeless works of classical literature. As a reader of a **TREDITION CLASSICS** book, the reader supports the mission to save many of the amazing works of world literature from oblivion.

The symbol of **TREDITION CLASSICS** is Johannes Gutenberg (1400 – 1468), the inventor of movable type printing.

With the series, tredition intends to make thousands of international literature classics available in printed format again – worldwide.

All books are available at book retailers worldwide in paperback and in hardcover. For more information please visit: www.tredition.com

tredition was established in 2006 by Sandra Latusseck and Soenke Schulz. Based in Hamburg, Germany, tredition offers publishing solutions to authors and publishing houses, combined with worldwide distribution of printed and digital book content. tredition is uniquely positioned to enable authors and publishing houses to create books on their own terms and without conventional manufacturing risks.

For more information please visit: www.tredition.com

Mendelism Third Edition

Reginald Crundall Punnett

Imprint

This book is part of the TREDITION CLASSICS series.

Author: Reginald Crundall Punnett
Cover design: toepferschumann, Berlin (Germany)

Publisher: tredition GmbH, Hamburg (Germany)
ISBN: 978-3-8491-5198-0

www.tredition.com
www.tredition.de

Copyright:
The content of this book is sourced from the public domain.

The intention of the TREDITION CLASSICS series is to make world literature in the public domain available in printed format. Literary enthusiasts and organizations worldwide have scanned and digitally edited the original texts. tredition has subsequently formatted and redesigned the content into a modern reading layout. Therefore, we cannot guarantee the exact reproduction of the original format of a particular historic edition. Please also note that no modifications have been made to the spelling, therefore it may differ from the orthography used today.

PREFACE

A few years ago I published a short sketch of Mendel's discovery in heredity, and of some of the recent experiments which had arisen from it. Since then progress in these studies has been rapid, and the present account, though bearing the same title, has been completely rewritten. A number of illustrations have been added, and here I may acknowledge my indebtedness to Miss Wheldale for the two coloured plates of sweet peas, to the Hon. Walter Rothschild for the butterflies figured on Plate VI., to Professor Wood for photographs of sheep, and to Dr. Drinkwater for the figures of human hands. To my former publishers also, Messrs. Bowes and Bowes, I wish to express my thanks for the courtesy with which they acquiesced in my desire that the present edition should be published elsewhere.

As the book is intended to appeal to a wide audience, I have not attempted to give more experimental instances than were necessary to illustrate the story, nor have I burdened it with bibliographical reference. The reader who desires further information may be referred to Mr. Bateson's indispensable Volume on *Mendel's Principles of Heredity* (Cambridge, 1909), where a full account of these matters is readily accessible. Neither have I alluded to recent cytological work in so far as it may bear upon our problems. Many of the facts connected with the division of the chromosomes are striking and suggestive, but while so much difference of opinion exists as to their interpretation they are hardly suited for popular treatment.

In choosing typical examples to illustrate the growth of our ideas it was natural that I should give the preference to those with which I was most familiar. For this reason the book is in some measure a record of the work accomplished by the Cambridge School of Genetics, and it is not unfair to say that under the leadership of William Bateson the contributions of this school have been second to none. But it should not be forgotten that workers in other European countries, and especially in America, have amassed a large and valuable body of evidence with which it is impossible to deal in a small volume of this scope.

It is not long since the English language was enriched by two new words—Eugenics and Genetics—and their similarity of origin has sometimes led to confusion between them on the part of those who are innocent of Greek. Genetics is the term applied to the experimental study of heredity and variation in animals and plants, and the main concern of its students is the establishing of law and order among the phenomena {vii}there encountered. Eugenics, on the other hand, deals with the improvement of the human race under existing conditions of law and sentiment. The Eugenist has to take into account the religious and social beliefs and prejudices of mankind. Other issues are involved besides the purely biological one, though as time goes on it is coming to be more clearly recognised that the Eugenic ideal is sharply circumscribed by the facts of heredity and variation, and by the laws which govern the transmission of qualities in living things. What these facts, what these laws are, in so far as we at present know them, I have endeavoured to indicate in the following pages; for I feel convinced that if the Eugenist is to achieve anything solid it is upon them that he must primarily build. Little enough material, it is true, exists at present, but that we now see to be largely a question of time and means. Whatever be the outcome, whatever the form of the structure which is eventually to emerge, we owe it first of all to Mendel that the foundations can be well and truly laid.

R. C. P.

Cambridge, March, 1911.

{ix}

CONTENTS

CHAPTER I
The Problem

CHAPTER II
Historical

CHAPTER III
Mendel's Work

CHAPTER IV
The Presence and Absence Theory

CHAPTER V
Interaction of Factors

CHAPTER VI
Reversion

CHAPTER VII
Dominance

{x} CHAPTER VIII
Wild Forms and Domestic Varieties

CHAPTER IX
Repulsion and Coupling of Factors

CHAPTER X

Sex

CHAPTER XI

Sex (*continued*)

CHAPTER XII

Intermediates

CHAPTER XIII

Variation and Evolution

CHAPTER XIV

Economical

CHAPTER XV

Man

APPENDIX

{xi}

ILLUSTRATIONS

PLATES

PLATE

I. Gregor Mendel
   Rabbits *To face*

II. Sweet Peas "

III. Sheep "

IV. Sweet Peas "

V. Fowls "

VI. Butterflies "

FIGURES IN TEXT

FIG.

1. Scheme of Inheritance in simple Mendelian Case
2. Feathers of Silky and Common Fowl
3. Single and Double Primulas
4. Fowls' Combs
5. Diagram of Inheritance of Fowls' Combs
6. Fowls' Combs
7. Diagram of F_2 Generation resulting from Cross between two White Sweet Peas
8. Diagram illustrating 9 : 3 : 4 Ratio in Mice
9. Sections of Primulas
10. Small and Large-eyed Primulas
11. Diagram illustrating Reversion in Pigeons
12. *Primula sinensis* × *Primula stellata*
13. Diagram illustrating Cross between Dominant and Recessive White Fowls
14. Bearded and Beardless Wheat

15. Feet of Fowls
16. Scheme of Inheritance of Horns in Sheep
17. *Abraxas grossulariata* and var. *lacticolor*
18. Scheme of Inheritance in *Abraxas*
19. Scheme of Inheritance of Silky Hen × Brown Leghorn Cock
20. Scheme of Inheritance of Brown Leghorn Hen × Silky Cock
21. Scheme of F_1 (ex Brown Leghorn × Silky Cock) crossed with pure Brown Leghorn
22. Scheme for Silky Hen × Brown Leghorn Cock
23. Scheme for Brown Leghorn Hen × Silky Cock
24. Diagram illustrating Nature of Offspring from Brown Leghorn Hen × F_1 Cock
25. Scheme to illustrate Heterozygous Nature of Brown Leghorn Hen
26. Scheme of Inheritance of Colour-blindness
27. Single and Double Stocks
28. F_2 Generation ex Silky Hen × Brown Leghorn Cock
29. Pedigree of Eurasian Family
30. Curve illustrating Influence of Selection

{xiii}

31. Curve illustrating Conception of pure Lines
32. Brachydactylous and Normal Hands
33. Radiograph of Brachydactylous Hand
34. Pedigree of Brachydactylous Family
35. Pedigree of Hæmophilic Family

{xiv}

For although it be a more new and difficult way, to find out the nature of things, by the things themselves; then by reading of Books, to take our knowledge upon trust from the opinions of Philosophers: yet must it needs be confessed, that the former is much more open, and lesse fraudulent, especially in the Secrets relating to *Natural Philosophy*.

>William Harvey,
>
>*Anatomical Exercitations*, 1653.

{1}

CHAPTER I

THE PROBLEM

A curious thing in the history of human thought so far as literature reveals it to us is the strange lack of interest shown in one of the most interesting of all human relationships. Few if any of the more primitive peoples seem to have attempted to define the part played by either parent in the formation of the offspring, or to have assigned peculiar powers of transmission to them, even in the vaguest way. For ages man must have been more or less consciously improving his domesticated races of animals and plants, yet it is not until the time of Aristotle that we have clear evidence of any hypothesis to account for these phenomena of heredity. The production of offspring by man was then held to be similar to the production of a crop from seed. The seed came from the man, the woman provided the soil. This remained the generally accepted view for many centuries, and it was not until the recognition of woman as more than a passive agent that the physical basis of heredity became established. That recognition was effected by the microscope, for only with its advent was actual {2}observation of the minute sexual cells made possible. After more than a hundred years of conflict lasting until the end of the eighteenth century, scientific men settled down to the view that each of the sexes makes a definite material contribution to the offspring produced by their joint efforts. Among animals the female contributes the ovum and the male the spermatozoon; among plants the corresponding cells are the ovules and pollen grains.

As a general rule it may be stated that the reproductive cells produced by the female are relatively large and without the power of independent movement. In addition to the actual living substance which is to take part in the formation of a new individual, the ova are more or less heavily loaded with the yolk substance that is to provide for the nutrition of the developing embryo during the early stages of its existence. The size of the ova varies enormously in different animals. In birds and reptiles where the contents of the egg form the sole resources of the developing young they are very large in comparison with the size of the animal which lays them. In mammals, on the other hand, where the young are parasitic upon the mother during the earlier stages of their growth, the eggs are minute and only contain the small amount of yolk that enables them to reach the stage at which they develop the processes for attaching themselves to the wall of the maternal uterus. But whatever the differences in the size and appearance of the ova produced by different {3}animals, they are all comparable in that each is a distinct and separate sexual cell which, as a rule, is unable to develop into a new individual of its species unless it is fertilised by union with a sexual cell produced by the male.

The male sexual cells are always of microscopic size and are produced in the generative gland or testis in exceedingly large numbers. In addition to their minuter size they differ from the ova in their power of active movement. Animals present various mechanisms by which the sexual elements may be brought into juxtaposition, but in all cases some distance must be traversed in a fluid or semifluid medium (frequently within the body of the female parent) before the necessary fusion can occur. To accomplish this latter end of its journey the spermatozoon is endowed with some form of motile apparatus, and this frequently takes the form of a long flagellum, or whip-like process, by the lashing of which the little creature propels itself much as a tadpole with its tail.

In plants as in animals the female cells or ovules are larger than the pollen grains, though the disparity in size is not nearly so marked. Still they are always relatively minute cells since the circumstances of their development as parasites upon the mother plant render it unnecessary for them to possess any great supply of food yolk. The ovules are found surrounded by maternal tissue in the ovary, but through the stigma and down the pistil a {4}potential passage is left for the male cell. The majority of flowers are hermaphrodite, and in many cases they are also self-fertilising. The anthers burst and the contained pollen grains are then shed upon the stigma. When this happens, the pollen cell slips through a little hole in its coat and bores its way down the pistil to reach an ovule in the ovary. Complete fusion occurs, and the minute embryo of a new plant immediately results. But for some time it is incapable of leading a separate existence, and, like the embryo mammal, it lives as a parasite upon its parent. By the parent it is provided with a protective wrapping, the seed coat, and beneath this the little embryo swells until it reaches a certain size, when as a ripe seed it severs its connection with the maternal organism. It is important to realise that the seed of a plant is not a sexual cell but a young individual which, except for the coat that it wears, belongs entirely to the next generation. It is with annual plants in some respects as with many butterflies. During one summer they are initiated by the union of two sexual cells and pass through certain stages of larval development—the butterfly as a caterpillar, the plant as a parasite upon its mother. As the summer draws to a close each passes into a resting-stage against the winter cold—the butterfly as a pupa and the plant as a seed, with the difference that while the caterpillar provides its own coat, that of the plant is provided by its mother. With the advent of spring both butterfly and {5}plant emerge, become mature, and themselves ripen germ cells which give rise to a new generation.

Whatever the details of development, one cardinal fact is clear. Except for the relatively rare instances of parthenogenesis a new individual, whether plant or animal, arises as the joint product of two sexual cells derived from individuals of different sexes. Such sexual cells, whether ovules or ova, spermatozoa or pollen grains, are known by the general term of **gametes**, or marrying cells, and the individual formed by the fusion or yoking together of two gametes is spoken of as a **zygote**. Since a zygote arises from the yoking together of two separate gametes, the individual so formed must be regarded throughout its life as a double structure in which the components brought in by each of the gametes remain intimately fused in a form of partnership. But when the zygote in its turn comes to form gametes, the partnership is broken and the process is reversed. The component parts of the dual structure are resolved, with the formation of a set of single structures, the gametes.

The life cycle of a species from among the higher plants or animals may be regarded as falling into three periods: (1) a period of isolation in the form of gametes, each a living unit incapable of further development without intimate association with another produced by the opposite sex; (2) a period of association in which two gametes become yoked together into a zygote and react upon one {6}another to give rise by a process of cell division to what we ordinarily term an individual with all its various attributes and properties; and (3) a period of dissociation when the single structured gametes separate out from that portion of the double structured zygote which constitutes its generative gland. What is the relation between gamete and zygote, between zygote and gamete? how are the properties of the zygote represented in the gamete, and in what manner are they distributed from the one to the other? — these are questions which serve to indicate the nature of the problem underlying the process of heredity.

Owing to their peculiar power of growth and the relatively large size to which they attain, many of the properties of zygotes are appreciable by observation. The colour of an animal or of a flower, the shape of a seed, or the pattern on the wings of a moth are all zygotic properties, and all capable of direct estimation. It is otherwise with the properties of gametes. While the difference between a black and a white fowl is sufficiently obvious, no one by inspection can tell the difference between the egg that will hatch into a black and that which will hatch into a white. Nor from a mass of pollen grains can any one to-day pick out those that will produce white from those that will produce coloured flowers. Nevertheless, we know that in spite of apparent similarity there must exist fundamental differences among the gametes, even {7}among those that spring from the same individual. At present our only way of appreciating those differences is to observe the properties of the zygotes which they form. And as it takes two gametes to form a zygote, we are in the position of attempting to decide the properties of two unknowns from one known. Fortunately the problem is not entirely one of simple mathematics. It can be attacked by the experimental method, and with what measure of success will appear in the following pages.

{8}

CHAPTER II

HISTORICAL

To Gregor Mendel, monk and abbot, belongs the credit of founding the modern science of heredity. Through him there was brought into these problems an entirely new idea, an entirely fresh conception of the nature of living things. Born in 1822 of Austro-Silesian parentage, he early entered the monastery of Brünn, and there in the seclusion of the cloister garden he carried out with the common pea the series of experiments which has since become so famous. In 1865 after eight years' work he published the results of his experiments in the *Proceedings of the Natural History Society of Brünn*, in a brief paper of some forty pages. But brief as it is the importance of the results and the lucidity of the exposition will always give it high rank among the classics of biological literature. For thirty-five years Mendel's paper remained unknown, and it was not until 1900 that it was simultaneously discovered by several distinguished botanists. The causes of this curious neglect are not altogether without interest. Hybridisation experiments before Mendel there had been in plenty. The classificatory work of {9}Linnaeus in the latter half of the eighteenth century had given a definite significance to the word species, and scientific men began to turn their attention to attempting to discover how species were related to one another. And one obvious way of attacking the problem was to cross different species together and see what happened. This was largely done during the earlier half of the nineteenth century, though such work was almost entirely confined to the botanists. Apart from the fact that plants lend themselves to hybridisation work more readily than animals, there was probably another reason why zoologists neglected this form of investigation. The field of zoology is a wider one than that of botany, presenting a far greater variety of type and structure. Partly owing to their importance in the study of medicine, and partly owing to their smaller numbers, the anatomy of the vegetable was far better known than that of the animal kingdom. It is, therefore, not surprising that the earlier part of the nineteenth century found the zoologists, under the influence of Cuvier and his pupils, devoting their entire energies to describing the anatomy of the new forms of animal life which careful search at home and fresh voyages of discovery abroad were continually bringing to light. During this period the zoologist had little inclination or inducement to carry on those investigations in hybridisation which were occupying the attention of some botanists. Nor did the efforts of the botanists afford much {10}encouragement to such work, for in spite of the labour devoted to these experiments, the results offered but a confused tangle of facts, contributing in no apparent way to the solution of the problem for which they had been undertaken. After half a century of experimental hybridisation the determination of the relation of species and varieties to one another seemed as remote as ever. Then in 1859 came the *Origin of Species*, in which Darwin presented to the world a consistent theory to account for the manner in which one species might have arisen from another by a process of gradual evolution. Briefly put, that theory was as follows: In any species of plant or animal the reproductive capacity tends to outrun the available food supply, and the resulting competition leads to an inevitable struggle for existence. Of all the individuals born, only a portion, and that often a very small one, can survive to produce offspring. According to Darwin's theory, the nature of the surviving portion is not determined by chance alone. No two individuals of a species are precisely alike, and among the variations that occur some enable their possessors to cope more successfully with the competitive conditions under which they exist. In comparison with their less favoured brethren they have a better chance of surviving in the struggle for existence and consequently of leaving offspring. The argument is completed by the further assumption of a principle of heredity, in virtue of which offspring tend to {11}resemble their parents more than other members of the species. Parents possessing a favourable variation tend to transmit that variation to their offspring, to some in greater, to others in less degree. Those possessing it in greater degree will again have a better chance of survival, and will transmit the favourable variation in even greater degree to some of their offspring. A competitive struggle for existence working in combination with certain principles of variation and heredity results in a slow and continuous transformation of species through the operation of a process which Darwin termed natural selection.

The coherence and simplicity of the theory, supported as it was by the great array of facts which Darwin had patiently marshalled together, rapidly gained the enthusiastic support of the great majority of biologists. The problem of the relation of species at last appeared to be solved, and for the next forty years zoologists and botanists were busily engaged in classifying by the light of Darwin's theory the great masses of anatomical facts which had already accumulated and in adding and classifying fresh ones. The study of comparative anatomy and embryology received a new stimulus, for with the acceptance of the theory of descent with modification it became incumbent upon the biologist to demonstrate the manner in which animals and plants differing widely in structure and appearance could be conceivably related to one another. Thenceforward the energies of both {12}botanists and zoologists have been devoted to the construction of hypothetical pedigrees suggesting the various tracks of evolution by which one group of animals or plants may have arisen from another through a long continued process of natural selection. The result of such work on the whole may be said to have shown that the diverse forms under which living things exist to-day, and have existed in the past so far as palaeontology can tell us, are consistent with the view that they are all related by the community of descent which the accepted theory of evolution demands, though as to the exact course of descent for any particular group of animals there is often considerable diversity of opinion. It is obvious that all this work has little or nothing to do with the manner in which species are formed. Indeed, the effect of Darwin's *Origin of Species* was to divert attention from the way in which species originate. At the time that it was put forward his explanation appeared so satisfying that biologists accepted the notions of variation and heredity there set forth and ceased to take any further interest in the work of the hybridisers. Had Mendel's paper appeared a dozen years earlier it is difficult to believe that it could have failed to attract the attention it deserved. Coming as it did a few years after the publication of Darwin's great work, it found men's minds set at rest on the problems that he raised and their thoughts and energies directed to other matters. {13}

Nevertheless one interesting and noteworthy attempt to give greater precision to the term heredity was made about this time. Francis Galton, a cousin of Darwin, working upon data relating to the breeding of Basset hounds, found that he could express on a definite statistical scheme the proportion in which the different colours appeared in successive generations. Every individual was conceived of as possessing a definite heritage which might be expressed as unity. Of this, ½ was on the average derived from the two parents (*i.e.* ¼ from each parent), ¼ from the four grandparents, ⅛ from the eight great-grandparents, and so on. *The Law of Ancestral Heredity*, as it was termed, expresses with fair accuracy some of the statistical phenomena relating to the transmission of characters in a mixed population. But the problem of the way in which characters are distributed from gamete to zygote and from zygote to gamete remained as before. Heredity is essentially a physiological problem, and though statistics may be suggestive in the initiation of experiment, it is upon the basis of experimental fact that progress must ultimately rest. For this reason, in spite of its ingenuity and originality, Galton's theory and the subsequent statistical work that has been founded upon it failed to give us any deeper insight into the nature of the hereditary process.

While Galton was working in England the German zoologist August Weismann was elaborating the complicated {14}theory of heredity which eventually appeared in his work on *The Germplasm* (1885), a book which will be remembered for one notable contribution to the subject. Until the publication of Weismann's work it had been generally accepted that the modifications brought about in the individual during its lifetime, through the varying conditions of nutrition and environment, could be transmitted to the offspring. In this biologists were but following Darwin, who held that the changes in the parent resulting from increased use or disuse of any part or organ were passed on to the children. Weismann's theory involved the conception of a sharp cleavage between the general body tissues or somatoplasm and the reproductive glands or germplasm. The individual was merely a carrier for the essential germplasm whose properties had been determined long before he was capable of leading a separate existence. As this conception ran counter to the possibility of the inheritance of "acquired characters," Weismann challenged the evidence upon which it rested and showed that it broke down wherever it was critically examined. By thus compelling biologists to revise their ideas as to the inherited effects of use and disuse, Weismann rendered a valuable service to the study of genetics and did much to clear the way for subsequent research.

A further important step was taken in 1895, when Bateson once more drew attention to the problem of the origin {15}of species, and questioned whether the accepted ideas of variation and heredity were after all in consonance with the facts. Speaking generally, species do not grade gradually from one to the other, but the differences between them are sharp and specific. Whence comes this prevalence of discontinuity if the process by which they have arisen is one of accumulation of minute and almost imperceptible differences? Why are not intermediates of all sorts more abundantly produced in nature than is actually known to be the case? Bateson saw that if we are ever to answer this question we must have more definite knowledge of the nature of variation and of the nature of the hereditary process by which these variations are transmitted. And the best way to obtain that knowledge was to let the dead alone and to return to the study of the living. It was true that the past record of experimental breeding had been mainly one of disappointment. It was true also that there was no tangible clue by which experiments might be directed in the present. Nevertheless in this kind of work alone there seemed any promise of ultimate success.

A few years later appeared the first volume of de Vries' remarkable book on *The Mutation Theory*. From a prolonged study of the evening primrose (*Oenothera*) de Vries concluded that new varieties suddenly arose from older ones by sudden sharp steps or mutations, and not by any process involving the gradual accumulation of minute {16}differences. The number of striking cases from among widely different plants which he was able to bring forward went far to convincing biologists that discontinuity in variation was a more widespread phenomenon than had hitherto been suspected, and not a few began to question whether the account of the mode of evolution so generally accepted for forty years was after all the true account. Such in brief was the outlook in the central problem of biology at the time of the rediscovery of Mendel's work.

{17}

CHAPTER III

MENDEL'S WORK

The task that Mendel set before himself was to gain some clear conception of the manner in which the definite and fixed varieties found within a species are related to one another, and he realised at the outset that the best chance of success lay in working with material of such a nature as to reduce the problem to its simplest terms. He decided that the plant with which he was to work must be normally self-fertilising and unlikely to be crossed through the interference of insects, while at the same time it must possess definite fixed varieties which bred true to type. In the common pea (*Pisum sativum*) he found the plant he sought. A hardy annual, prolific, easily worked, *Pisum* has a further advantage in that the insects which normally visit flowers are unable to gather pollen from it and so to bring about cross fertilisation. At the same time it exists in a number of strains presenting well-marked and fixed differences. The flowers may be purple, or red, or white; the plants may be tall or dwarf; the ripe seeds may be yellow or green, round or wrinkled — such are a few of the characters in which the various races of peas differ from one another. {18}

In planning his crossing experiments Mendel adopted an attitude which marked him off sharply from the earlier hybridisers. He realised that their failure to elucidate any general principle of heredity from the results of cross fertilisation was due to their not having concentrated upon particular characters or traced them carefully through a sequence of generations. That source of failure he was careful to avoid, and throughout his experiments he crossed plants presenting sharply contrasted characters, and devoted his efforts to observing the behaviour of these characters in successive generations. Thus in one series of experiments he concentrated his attention on the transmission of the characters tallness and dwarfness, neglecting in so far as these experiments were concerned any other characters in which the parent plants might differ from one another. For this purpose he chose two strains of peas, one of about 6 feet in height, and another of about 1½ feet. Previous testing had shown that each strain bred true to its peculiar height. These two strains were artificially crossed [1] with one another, and it was found to make no difference which was used as the pollen parent and which was used as the ovule parent. In either case the result was the same. The result of crossing tall with dwarf was in every case nothing but talls, as tall or even a little taller than the tall parent. For this reason Mendel termed tallness the **dominant** and {19}dwarfness the **recessive** character. The next stage was to collect and sow the seeds of these tall hybrids. Such seeds in the following year gave rise to a mixed generation consisting of talls and dwarfs *but no intermediates*. By raising a considerable number of such plants Mendel was able to establish the fact that the number of talls which occurred in this generation was almost exactly three times as great as the number of the dwarfs. As in the previous year, seed were carefully collected from this, the second hybrid generation, and in every case *the seeds from each individual plant were harvested separately and separately sown in the following year*. By this respect for the individuality of the different plants, however closely they resembled one another, Mendel found the clue that had eluded the efforts of all his predecessors. The seeds collected from the dwarf recessives bred true, giving nothing but dwarfs. And this was true for every dwarf tested. But with the talls it was quite otherwise. Although indistinguishable in appearance, some of them bred true, while others behaved like the original tall hybrids, giving a generation consisting of talls and dwarfs in the proportion of three of {20}the former to one of the latter. Counting showed that the number of the talls which gave dwarfs was double that of the talls which bred true.

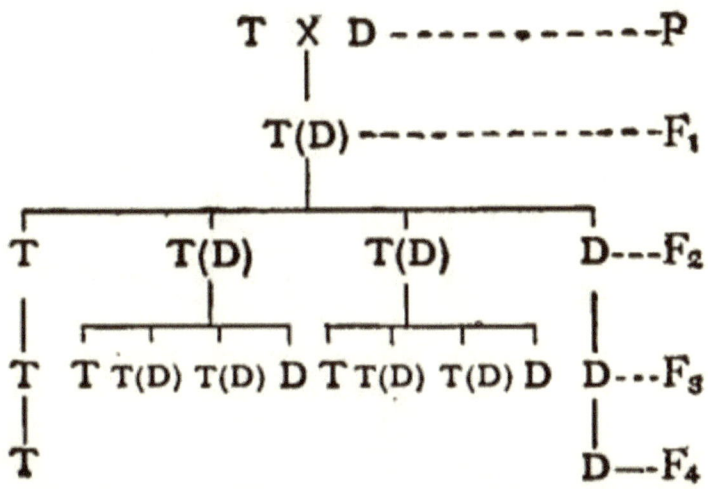

If we denote a dwarf plant as D, a true breeding tall plant as T, and a tall which gives both talls and dwarfs in the ratio 3 : 1 as T(D), the result of these experiments may be briefly summarised in the foregoing scheme. [2]

Mendel experimented with other pairs of contrasted characters and found that in every instance they followed the same scheme of inheritance. Thus coloured flowers were dominant to white, in the ripe seeds yellow was dominant to green, and round shape was dominant to wrinkled, and so on. In every case where the inheritance of an alternative pair of characters was concerned the effect of the cross in successive generations was to produce three and only three different sorts of individuals, viz. dominants which bred true, dominants which gave both dominant and recessive offspring in the ratio 3 : 1, and recessives which always bred true. Having determined a general scheme of inheritance which experiment showed to hold good for each of the seven pairs of alternative characters with which he worked, Mendel set himself to providing a theoretical interpretation of this scheme which, as he clearly realised, must be in terms of germ cells. He {21}conceived of the gametes as bearers of something capable of giving rise to the characters of the plant, but he regarded any individual gamete as being able to carry one and one only of any alternative pair of characters. A given gamete could carry tallness *or* dwarfness, but not both. The two were mutually exclusive so far as the gamete was concerned. It must be pure for one or the other of such a pair, and this conception of the purity of the gametes is the most essential part of Mendel's theory.

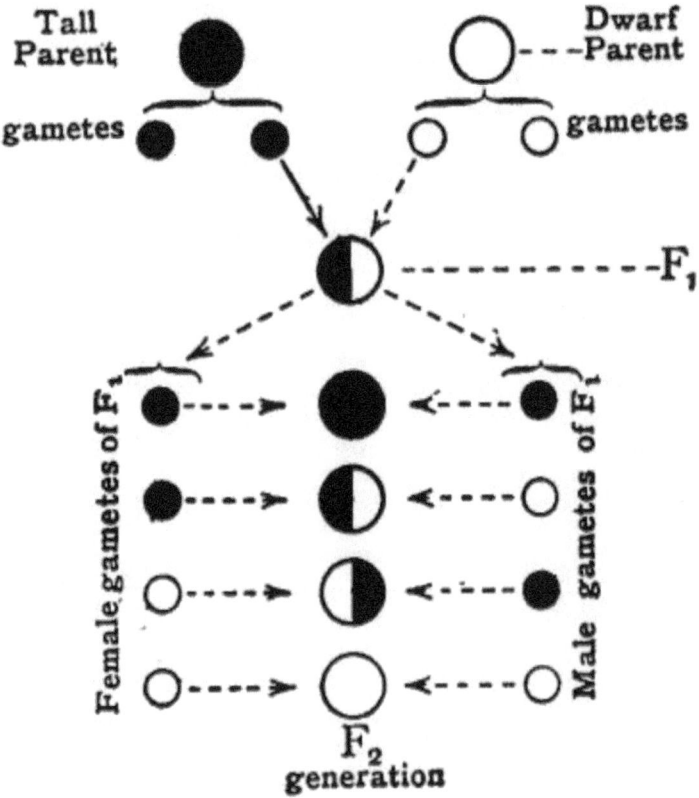

Fig. 1.

Scheme of inheritance in the cross of tall with dwarf pea. Gametes represented by small and zygotes by larger circles.

We may now proceed with the help of the accompanying scheme (Fig. 1) to deduce the results that should flow from Mendel's conception of the nature of the gametes, and to see how far they are in accordance with the facts. Since the original tall plant belonged to a strain which bred true, all the gametes produced by it must bear the tall character. Similarly all the gametes of the original dwarf plant must bear the dwarf character. A cross between these two means the union of {22}a gamete containing tallness with one bearing dwarfness. Owing to the completely dominant nature of the tall character, such a plant is in appearance indistinguishable from the pure tall, but it differs markedly from it in the nature of the gametes to which it gives rise. When the formation of the gametes occurs, the elements representing dwarfness and tallness **segregate** from one another, so that half of the gametes produced contain the one, and half contain the other of these two elements. For on hypothesis every gamete must be pure for one or other of these two characters. And this is true for the ovules as well as for the pollen grains. Such hybrid F_1 plants, therefore, must produce a series of ovules consisting of those bearing tallness and those bearing dwarfness, and must produce them in equal numbers. And similarly for the pollen grains. We may now calculate what should happen when such a series of pollen grains meets such a series of ovules, *i.e.* the nature of the generation that should be produced when the hybrid is allowed to fertilise itself. Let us suppose that there are $4x$ ovules so that $2x$ are "tall" and $2x$ are "dwarf." These are brought in contact with a mass of pollen grains of which half are "tall" and half are "dwarf." It is obvious that a "tall" ovule has an equal chance of being fertilised by a "tall" or a "dwarf" pollen grain. Hence of our $2x$ "tall" ovules, x will be fertilised by "tall" pollen grains and x will be fertilised by "dwarf" pollen grains. The former must give rise to tall {23}plants, and since the dwarf character has been entirely eliminated from them they must in the future breed true. The latter must also give rise to tall plants, but since they carry also the recessive dwarf character they must when bred from produce both tails and dwarfs. Each of the $2x$ dwarf ovules, again, has an equal chance of being fertilised by a "tall" or by a "dwarf" pollen grain. Hence x will give rise to tall plants carrying the recessive dwarf character, while x will produce plants from which the tall character has been eliminated, *i.e.* to pure recessive dwarfs. Consequently from the $4x$ ovules of the self-fertilised hybrid we ought to obtain $3x$ tall and x dwarf plants. And of the $3x$ talls x should breed true to tallness, while the remaining $2x$, having been formed like the original hybrid by the union of a "tall" and a "dwarf" gamete, ought to behave like it when bred from and give talls and dwarfs in the ratio 3 : 1. Now this is precisely the result actually obtained by experiment (cf. p. 17), and the close accord of the experimental results with those deduced on the assumption of the purity of the gametes as enunciated by Mendel affords the strongest of arguments for regarding the nature of the gametes and their relation to the characters of the zygotes in the way that he has done.

It is possible to put the theory to a further test. The explanation of the 3 : 1 ratio of dominants and recessives in the F_2 generation is regarded as due to the F_1 individuals producing equal numbers of gametes bearing the {24}dominant and recessive elements respectively. If now the F_1 plant be crossed with the pure recessive, we are bringing together a series of gametes consisting of equal numbers of dominants and recessives with a series consisting solely of recessives. We ought from such a cross to obtain equal numbers of dominant and recessive individuals, and further, the dominants so produced ought all to give both dominants and recessives in the ratio 3 : 1 when they themselves are bred from. Both of these expectations were amply confirmed by experiment, and crossing with the recessive is now a recognised way of testing whether a plant or animal bearing a dominant character is a pure dominant, or an impure dominant which is carrying the recessive character. In the former case the offspring will be all of the dominant form, while in the latter they will consist on the average of equal numbers of dominants and recessives.

So far we have been concerned with the results obtained when two individuals differing in a single pair of characters are crossed together and with the interpretation of those results. But Mendel also used plants which differed in more than a single pair of differentiating characters. In such cases he found that each pair of characters followed the same definite rule, but that the inheritance of each pair was absolutely independent of the other. Thus, for example, when a tall plant bearing coloured flowers was crossed with a dwarf plant {25}bearing white flowers the resulting hybrid was a tall plant with coloured flowers. For coloured flowers are dominant to white, and tallness is dominant to dwarfness. In the succeeding generation there are plants with coloured flowers and plants with white flowers in the proportion of 3 : 1, and at the same time tall plants and dwarf plants in the same proportion. Hence the chances that a tall plant will have coloured flowers are three times as great as its chance of having white flowers. And this is also true for the dwarf plants. As the result of this cross, therefore, we should expect an F_2 generation consisting of four classes, viz. coloured talls, white talls, coloured dwarfs, and white dwarfs, and we should further expect these four forms to appear in the ratio of 9 coloured talls, 3 white talls, 3 coloured dwarfs, and 1 white dwarf. For this is the only ratio which satisfies the conditions that the talls should be to the dwarfs as 3 : 1, and at the same time the coloured should be to the whites as 3 : 1. And these are the proportions that Mendel found to obtain actually in his experiments. Put in a more general form, it may be stated that when two individuals are crossed which differ in two pairs of differentiating characters the hybrids (F_1) are all of the same form, exhibiting the dominant character of each of the two pairs, while the F_2 generation produced by such hybrids consists on the average of 9 showing both dominants, 3 showing one dominant and one recessive, {26}3 showing the other dominant and the other recessive, and 1 showing both recessive characters. And, as Mendel pointed out, the principle may be extended indefinitely. If, for example, the parents differ in three pair of characters A, B, and C, respectively dominant to a, b, and c, the F_1 individuals will be all of the form ABC, while the F_2 generation will consists of 27 ABC, 9 ABc, 9 AbC, 9 aBC, 3 Abc, 3 aBc, 3 abC, and 1 abc. When individuals differing in a number of alternative characters are crossed together, the hybrid generation, provided that the original parents were of pure strains, consists of plants of the same form; but when these are bred from a redistribution of the various characters occurs. That redistribution follows the same definite rule for each character, and if the constitution of the original parents be known, the nature of the F_2 generation, *i.e.* the number of possible forms and the proportions in which they occur, can be readily calculated. Moreover, as Mendel showed, we can calculate also the chances of any given form breeding true. To this point, however, we shall return later.

Of Mendel's experiments with beans it is sufficient to say here that they corroborated his more ample work with peas. He is also known to have made experiments with many other plants, and a few of his results are incidentally given in his series of letters to Nägeli the botanist. To the breeding and crossing of bees he also devoted much {27}time and attention, but unhappily the record of these experiments appears to have been lost. The only other published work that we possess dealing with heredity is a brief paper on some crossing experiments with the hawkweeds (*Hieracium*), a genus that he chose for working with because of the enormous number of forms under which it naturally exists. By crossing together the more distinct varieties, he evidently hoped to produce some of these numerous wild forms, and so throw light upon their origin and nature. In this hope he was disappointed. Owing in part to the great technical difficulties attending the cross fertilisation of these flowers he succeeded in obtaining very few hybrids. Moreover, the behaviour of those which he did obtain was quite contrary to what he had found in the peas. Instead of giving a variety of forms in the F_2 generation, they bred true and continued to do so as long as they were kept under observation. More recent research has shown that this is due to a peculiar form of parthenogenesis (cf. p. 135), and not to any failure of the characters to separate clearly from one another in the gametes. Mendel, however, could not have known of this, and his inability to discover in *Hieracium* any indication of the rule which he had found to hold good for both peas and beans must have been a source of considerable disappointment. Whether for this reason, or owing to the utter neglect of his work by the scientific world, Mendel gave up his experimental {28}researches during the latter part of his life. His closing years were shadowed with ill-health and embittered by a controversy with the Government on a question of the rights of his monastery. He died of Bright's disease in 1884.

> *Note.*—Shortly after the discovery of Mendel's paper a need was felt for terms of a general nature to express the constitution of individuals in respect of inherited characters, and Bateson accordingly proposed the words homozygote and heterozygote. An individual is said to be homozygous for a given character when it has been formed by two gametes each bearing the character, and all the gametes of a homozygote bear the character in re-

spect of which it is homozygous. When, however, the zygote is formed by two gametes of which one bears the given character while the other does not, it is said to be heterozygous for the character in question, and only half the gametes produced by such a heterozygote bear the character. An individual may be homozygous for one or more characters, and at the same time may be heterozygous for others.

{29}

CHAPTER IV

THE PRESENCE AND ABSENCE THEORY

It was fortunate for the development of biological science that the rediscovery of Mendel's work found a small group of biologists deeply interested in the problems of heredity, and themselves engaged in experimental breeding. To these men the extraordinary significance of the discovery was at once apparent. From their experiments, undertaken in ignorance of Mendel's paper, de Vries, Correns, and Tschermak were able to confirm his results in peas and other plants, while Bateson was the first to demonstrate their application to animals. Thenceforward the record has been one of steady progress, and the result of ten years' work has been to establish more and more firmly the fundamental nature of Mendel's discovery. The scheme of inheritance, which he was the first to enunciate, has been found to hold good for such diverse things as height, hairiness, and flower colour and flower form in plants, the shape of pollen grains, and the structure of fruits; while among animals the coat colour of mammals, the form of the feathers and of the comb in poultry, the waltzing habit of Japanese mice, and eye {30}colour in man are but a few examples of the diversity of characters which all follow the same law of transmission. And as time went on many cases which at first seemed to fall without the scheme have been gradually brought into line in the light of fuller knowledge. Some of these will be dealt with in the succeeding chapters of this book. Meanwhile we may concern ourselves with the single modification of Mendel's original views which has arisen out of more ample knowledge.

Fig. 2.

A wing feather and a contour feather of an ordinary and a silky fowl. The peculiar ragged appearance of the silky feathers is due to the absence of the little hooks or barbules which hold the barbs together. The silky condition is recessive.

As we have already seen, Mendel considered that in the gamete there was either a definite something {31}corresponding to the dominant character or a definite something corresponding to the recessive character, and that these somethings whatever they were could not coexist in any single gamete. For these somethings we shall in future use the term **factor**. The factor, then, is what corresponds in the gamete to the **unit-character** that appears in some shape or other in the development of the zygote. Tallness in the pea is a unit-character, and the gametes in which it is represented are said to contain the factor for tallness. Beyond their existence in the gamete and their mode of transmission we make no suggestion as to the nature of these factors.

Fig. 3.

Two double and an ordinary single primula flower. This form of double is recessive to the single.

{32}

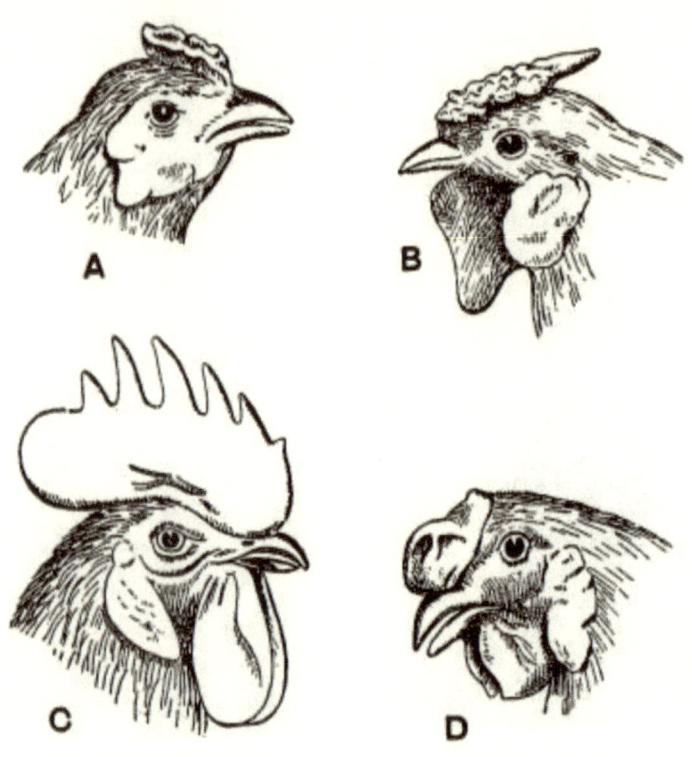

Fig. 4.

Fowls' combs. A, pea; B, rose; C, single; D, walnut.

On Mendel's view there was a factor corresponding to the dominant character and another factor corresponding to the recessive character of each alternative pair of unit-characters, and the characters were alternative because no gamete could carry more than one of the two factors belonging to the alternative pair. On the other hand, Mendel supposed that it always carried either one or the other of such a pair. As experimental work proceeded, {33}it soon became clear that there were cases which could not be expressed in terms of this conception. The nature of the difficulty and the way in which it was met will perhaps be best understood by considering a set of experiments in which it occurred. Many of the different breeds of poultry are characterised by a particular form of comb, and in certain cases the inheritance of these has been carefully worked out. It was shown that the rose comb (Fig. 4, B) with its flattened papillated upper surface and backwardly projecting pike was dominant in the ordinary way to the deeply serrated high single comb (Fig. 4, C) which is characteristic of the Mediterranean races. Experiment also showed that the pea comb (Fig. 4, A), a form with a low central and two well-developed lateral ridges, such as is found in Indian game, behaves as a simple dominant to the single comb. The interesting question arose as to what would happen when the rose and the pea, two forms each dominant to the same third form, were mated together. It seemed reasonable to suppose that things which were alternative to the same thing would be alternative to one another — that either rose or pea would dominate in the hybrids, and that the F_2 generation would consist of dominants and recessives in the ratio 3 : 1. The result of the experiment was, however, very different. The cross rose × pea led to the production of a comb quite unlike either of them. This, the so-called walnut comb (Fig. 4, D), {34}from its resemblance to the half of a walnut, is a type of comb which is normally characteristic of the Malay fowl. Moreover, when these F_1 birds were bred together, a further unlooked-for result was obtained. As was expected, there appeared in the F_2 generation the three forms walnut, rose, and pea. But there also appeared a definite proportion of single-combed birds, and among many hundreds of chickens bred in this way the proportions in which the four forms walnut, rose, pea, and single appeared was 9 : 3 : 3 : 1.

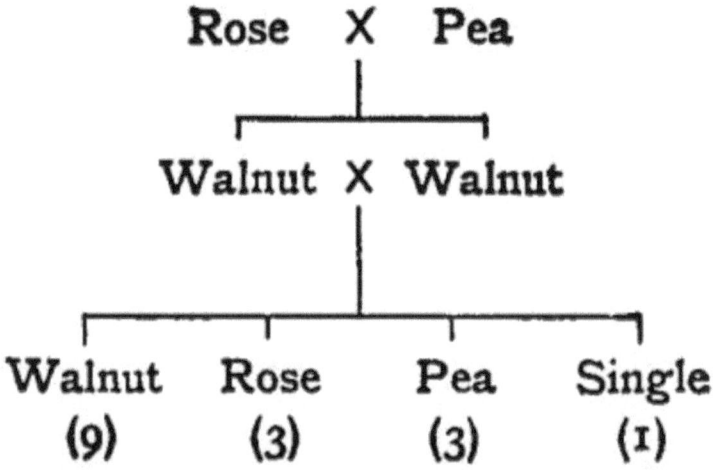

Now this, as Mendel showed, is the ratio found in an F_2 generation when the original parents differ in two pairs of alternative characters, and from the proportions in which the different forms of comb occur we must infer that the walnut contains both dominants, the rose and the pea one dominant each, while the single is pure for both recessive characters. This accorded with subsequent breeding experiments, for the singles bred perfectly true as soon as they had once made their appearance. So far the case is clear. The difficulty comes when we attempt to define these two pairs of characters. How are we to express the fact that while single behaves as a simple recessive to either pure rose, or to pure pea, it can yet appear in F_2 from a cross {35}between these two pure forms, though neither of them should, on Mendel's view, contain the single? An explanation which covers the facts in a simple way is that which has been termed the "Presence and Absence" theory. On this theory the dominant character of an alternative pair owes its dominance to the presence of a factor which is absent in the recessive. The tall pea is tall owing to the presence in it of the factor for tallness, but in the absence of this factor the pea remains a dwarf. All peas are dwarf, but the tall is a dwarf plus a factor which turns it into a tall. Instead of the characters of an alternative pair being due to two separate factors, we now regard them as the expression of the only two possible states of a single factor, viz. its presence or its absence. The conception will probably become clearer if we follow its application in detail to the case of the fowl's combs. In this case we are concerned with the transmission of the two factors, rose (R) and pea (P), the presence of each of which is alternative to its absence. The rose-combed bird contains the factor for rose but not that for pea, and the pea-combed bird contains the factor for pea but not that for rose. When both factors are present in a bird, as in the hybrid made by crossing rose with pea, the result is a walnut. For convenience of argument we may denote the presence of a given factor by a capital letter and its absence by the corresponding small letter. The use of the small letter is merely a symbolic way of intimating {36}that a particular factor is absent in a gamete or zygote. Represented thus the zygotic constitution of a pure rose-combed bird is $RRpp$; for it has been formed by the union of two gametes both of which contained R but not P. Similarly we may denote the pure pea-combed bird as $rrPP$. On crossing the rose with the pea union occurs between a gamete Rp and a gamete rP, resulting in the formation of a heterozygote of the constitution $RrPp$. The use of the small letters here informs us that such a zygote contains only a single dose of each of the factors R and P, although, of course, it is possible for a zygote, if made in a suitable way, to have a double dose of any factor. Now when such a bird comes to form gametes a separation takes place between the part of the zygotic cell containing R and the part which does not contain it (r). Half of its gametes, therefore, will contain R and the other half will be without it (r). Similarly half of its gametes will contain P and the other half will be without it (p). It is obvious that the chances of R being distributed to a gamete with or without P are equal. Hence the gametes containing R will be of two sorts, RP and Rp, and these will be produced in equal numbers. Similarly the gametes without R will also be of two sorts, rP and rp, and these, again, will be produced in equal numbers. Each of the hybrid walnut-combed birds, therefore, gives rise to a series consisting of equal numbers of gametes of the four different types RP, Rp, rP, and rp; and the breeding {37}together of such F_1 birds means the bringing together of two such series of gametes. When this happens an ovum of any one of the four types has an equal chance of being fertilised by a spermatozoon of any one of the four types. A convenient and simple method of demonstrating what happens under such circumstances is the method sometimes termed the "chessboard" method. For two series each consisting of four different types of gamete we require a square divided up into 16 parts. The four terms of the gametic series are first written horizontally across the four sets of four squares, so that the series is repeated four times. It is then written vertically four times, care being taken to keep to the same order. In this simple mechanical way all the possible combinations are represented and in their proper proportions.

RP RP Walnut	RP Rp Walnut	RP rP Walnut	RP rp Walnut
Rp RP Walnut	Rp Rp Rose	Rp rP Walnut	Rp rp Rose
rP RP Walnut	rP Rp Walnut	rP rP Pea	rP rp Pea
rp RP Walnut	rp Rp Rose	rp rP Pea	rp rp Single

Fig. 5.
Diagram to illustrate the nature of the F_2 generation from the cross of rose comb × pea comb. Fig. 5 shows the result of applying this method to our series *RP*, *Rp*, *rP*, *rp*, and the 16 squares represent the different kinds of zygotes formed and the proportions in which they occur. As {38}the figure shows, 9 zygotes contain both *R* and *P*, having a double or a single dose of either or both of these factors.

Such birds must be all walnut combed. Three out of the 16 zygotes contain R but not P, and these must be rose-combed birds. Three, again, contain P but not R and must be pea-combed birds. Finally one out of the 16 contains neither R nor P. It cannot be rose — it cannot be pea. It must, therefore, be something else. As a matter of fact it is single. Why it should be single and not something else follows from what we already know about the behaviour of these various forms of comb. For rose is dominant to single; therefore on the Presence and Absence theory a rose is a single plus a factor which turns the single into a rose. If we could remove the "rose" factor from a rose-combed bird the underlying single would come into view. Similarly a pea comb is a single plus a factor which turns the single into a pea, and a walnut is a single which possesses two additional modifying factors. Singleness, in fact, underlies all these combs, and if we write their zygotic constitution in full we must denote a walnut as *RRPPSS*, a rose as *RRppSS*, a pea as *rrPPSS*, and a single as *rrppSS*. The crossing of rose with pea results in a reshuffling of the factors concerned, and in accordance with the principle of segregation some zygotes are formed in which neither of the modifying factors R and P are present, and the single character can then become manifest. {39}

The Presence and Absence theory is to-day generally accepted by students of these matters. Not only does it afford a simple explanation of the remarkable fact that in all cases of Mendelian inheritance we should be able to express our unit-characters in terms of alternative pairs, but, as we shall have occasion to refer to later, it suggests a clue as to the course by which the various domesticated varieties of plants and animals have arisen from their wild prototypes.

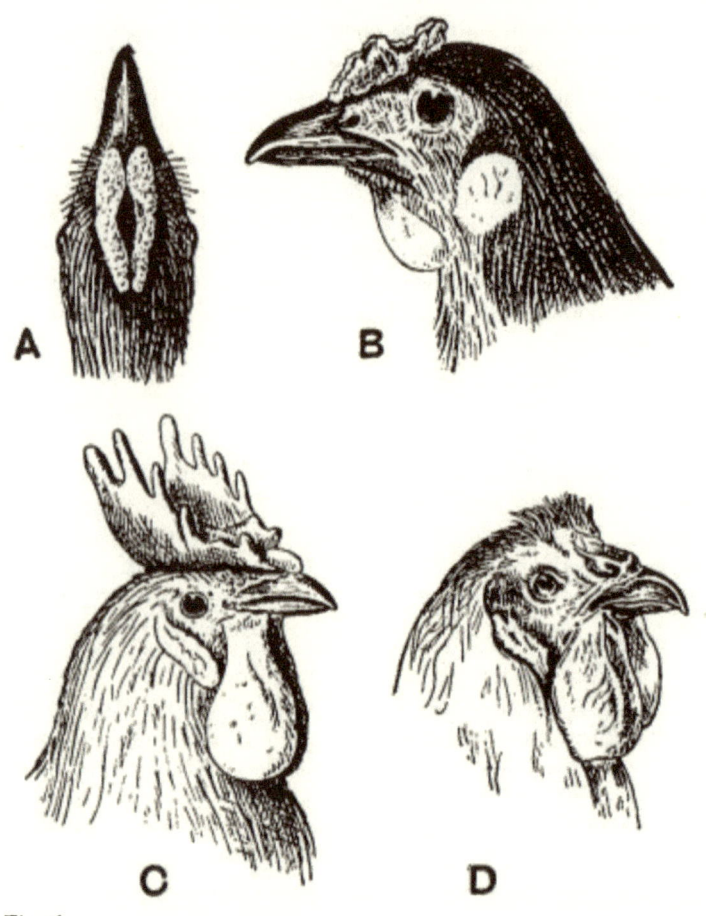

Fig. 6.

Fowls' combs. A and B, F_1 hen from rose × Breda; C, an F_1 cock from the cross of single × Breda; D, head of Breda cock.

Before leaving this topic we may draw attention to some experiments which offer a pretty confirmation of the view that the rose comb is a single to which a modifying factor for roseness has been added. It was argued that if we could find a type of comb in which the factor for singleness was absent, then on crossing such a comb with a rose we ought, if singleness really underlies rose, to obtain some single combs in F$_2$ from such a cross. Such a comb we had the good fortune to find in the Breda fowl, a breed largely used in Holland. This fowl is usually spoken of as combless, for the place of the comb is taken by a covering of short bristlelike feathers (Fig. 6, D). In reality it possesses the vestige of a comb in the form of two minute lateral knobs of comb tissue. Characteristic also of this breed is the high development of the horny nostrils, a feature probably correlated with the almost complete absence of comb. The first step in the experiment was to prove the absence of the factor for singleness in the Breda. {40}On crossing Breda with single the F$_1$ birds exhibit a large comb of the form of a double single comb in which the two portions are united anteriorly, but diverge from one another towards the back of the head (Fig. 6, C). The Breda contains an element of duplicity which is dominant to the simplicity of the ordinary single comb. But it cannot contain the factor for the single comb, because as soon as that is put into it by crossing with a single the comb {41}

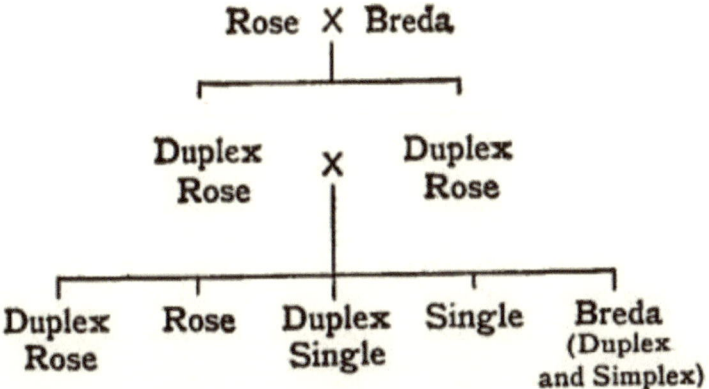

assumes a large size, and is totally distinct in appearance from its almost complete absence in the pure Breda. Now when the Breda is crossed with the rose duplicity is dominant to simplicity, and rose is dominant to lack of comb, and the F$_1$ generation consists of birds possessing duplex rose combs (Fig. 6, A and B). On breeding such birds together we obtain a generation consisting of Bredas, duplex roses, roses, duplex singles, and singles. From our previous experiment we know that the singles could not have come from the Breda, since a Breda comb to which the factor for single has been added no longer remains a Breda. Therefore it must have come from the rose, thus confirming our view that the rose is in reality a single comb which contains in addition a dominant modifying factor (R) whose presence turns it into a rose. We shall take it, therefore, that there is good experimental evidence for the Presence and Absence theory, and we shall express in terms of it the various cases which come up for discussion in succeeding chapters.

{42}

CHAPTER V

INTERACTION OF FACTORS

We have now reached a point at which it is possible to formulate a definite conception of the living organism. A plant or animal is a living entity whose properties may in large measure be expressed in terms of unit-characters, and it is the possession of a greater or lesser number of such unit-characters renders it possible for us to draw sharp distinctions between one individual and another. These unit-characters are represented by definite factors in the gamete which in the process of heredity behave as indivisible entities, and are distributed according to a definite scheme. The factor for this or that unit-character is either present in the gamete or it is not present. It must be there in its entirety or completely absent. Such at any rate is the view to which recent experiment has led us. But as to the nature of these factors, the conditions under which they exist in the gamete, and the manner in which they produce their specific effects in the zygote, we are at present almost completely in the dark.

The case of the fowls' combs opens up the important question of the extent to which the various factors can {43}influence one another in the zygote. The rose and the pea factors are separate entities, and each when present alone produces a perfectly distinct and characteristic effect upon the single comb, turning it into a rose or a pea as the case may be. But when both are present in the same zygote their combined effect is to produce the walnut comb, a comb which is quite distinct from either and in no sense intermediate between them. The question of the influence of factors upon one another did not present itself to Mendel because he worked with characters which affected different parts of the plant. It was unlikely that the factor which led to the production of colour in the flower would affect the shape of the pod, or that the height of the plant would be influenced by the presence or absence of the factor that determined the shape of the ripe seed. But when several factors can modify the same structure it is reasonable to suppose that they will influence one another in the effects which their simultaneous presence has upon the zygote. By themselves the pea and the rose factors each produce a definite modification of the single comb, but when both are present in the zygote, whether as a single or double dose, the modification that results is quite different to that produced by either when present alone. Thus we are led to the conception of characters which depend for their manifestation on more than one factor in the zygote, and in the present chapter we may consider a few of the {44}phenomena which result from such interaction between separate and distinct factors.

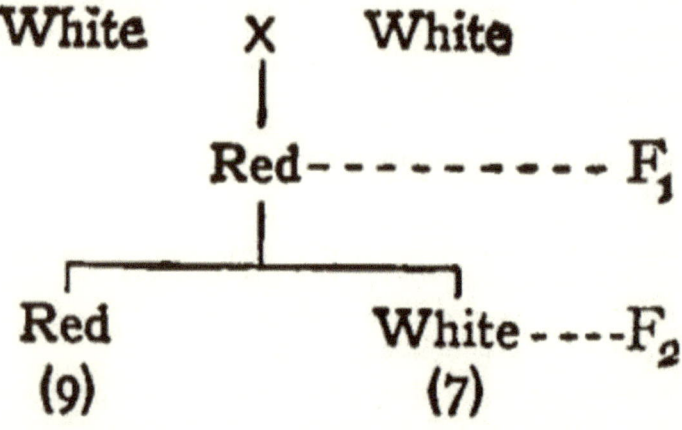

One of the most interesting and instructive cases in which the interaction between separate factors has been demonstrated is a case in the sweet pea. All white sweet peas breed true to whiteness. And generally speaking the result of crossing different whites is to produce nothing but whites, whether in F_1 or in succeeding generations. But there are certain strains of white sweet peas which when crossed together produce only coloured flowers. The colour may be different in different cases, though for our present purpose we may take a case in which the colour is red. When such reds are allowed to self-fertilise themselves in the normal way and the seeds sown, the resulting F_2 generation consists of reds and whites, the former being rather more numerous than the latter in the proportion of 9 : 7. The raising of a further generation from the seeds of these F_2 plants shows that the whites always breed true to whiteness, but that different reds may behave differently. Some breed true, others give reds and whites in the ratio 3 : 1, while others, again, give reds and whites in the ratio 9 : 7. As in the case of the fowls' combs, this case may be interpreted in terms of the presence and absence of two factors. {45}

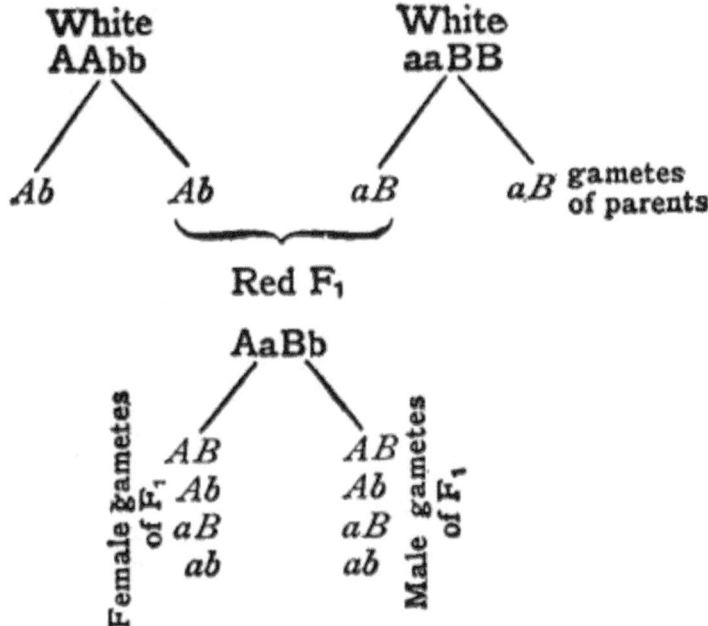

Red in the sweet pea results from the interaction of two factors, and unless these are both present the red colour cannot appear. Each of the white parents carried one of the two factors whose interaction is necessary for the production of the red colour, and as a cross between them brings these two complementary factors together the F_1 plants must all be red. As this case is of considerable importance for the proper understanding of much that is to follow, and as it has been completely worked out, we shall consider it in some detail. Denoting these two colour factors by *A* and *B* respectively we may proceed to follow out the consequences of this cross. Since all the F_1 plants were red the constitution of the parental whites must have been *AAbb* and *aaBB* respectively, and their gametes consequently *Ab* and *aB*. The constitution of the F_1 plants must, therefore, be *AaBb*. Such a plant being heterozygous for two factors produces a series of gametes of the four kinds *AB*, *Ab*, *aB*, *ab*, and produces them in equal numbers (cf. p. 36). To obtain the various types of zygotes which are produced when such {46}

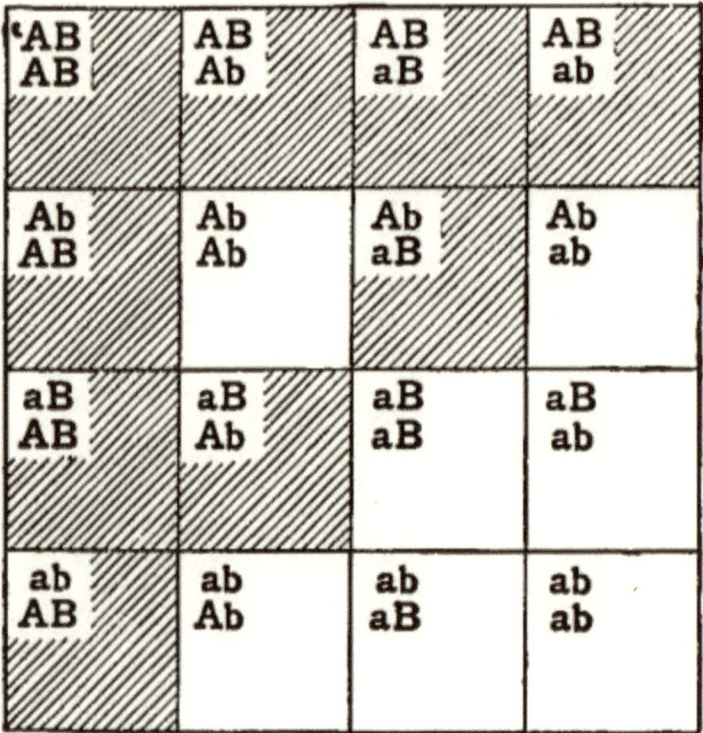

Fig. 7.
Diagram to illustrate the nature of the F_2 generation from the two white sweet peas which give a coloured F_1. a series of pollen grains meets a similar series of ovules we may make use of the same "chessboard" system which we have already adopted in the case of the fowls' combs. An examination of this figure (Fig. 7) shows that 9 out of the 16 squares contain both *A* and *B*, while 7 contain either *A* or *B* alone, or neither. In other words, on this view of the nature of the two white sweet peas we should in the F_2 generation look for the appearance of coloured and white flowers in the ratio 9 : 7. And this, as we have already seen, is what was actually found by experiment. Further examination of the figure shows that the coloured plants are not all of the same constitution, but are of four kinds with

respect to their zygotic constitution, viz. *AABB*, *AABb*, *AaBB*, and *AaBb*. Since *AABB* is homozygous for both *A* and *B*, all the gametes which it produces must contain both of these factors, and such a plant must therefore breed true to the red colour. A plant of the {47}constitution *AABb* is homozygous for the factor *A*, but heterozygous for *B*. All of its gametes will contain *A*, but only one-half of them will contain *B*, *i.e.* it produces equal numbers of gametes *AB* and *Ab*. Two such series of gametes coming together must give a generation consisting of x *AABB*, $2x$ *AABb*, and x *AAbb*, that is, reds and whites in the ratio 3 : 1. Lastly the red zygotes of the constitution *AaBb* have the same constitution as the original red made from the two whites, and must therefore when bred from give reds and whites in the ratio 9 : 7. The existence of all these three sorts of reds was demonstrated by experiment, and the proportions in which they were met with tallied with the theoretical explanation.

The theory was further tested by an examination into the properties of the various F_2 whites which come from a coloured plant that has itself been produced by the mating of two whites. As Fig. 7 shows, these are, in respect of their constitution, of five different kinds, viz. *AAbb*, *Aabb*, *aaBB*, *aaBb*, and *aabb*. Since none of them produce anything but whites on self-fertilisation it was found necessary to test their properties in another way, and the method adopted was that of crossing them together. It is obvious that when this is done we should expect different results in different cases. Thus the cross between two whites of the constitution *AAbb* and *aaBB* should give nothing but coloured plants; for these two whites are of {48}the same constitution as the original two whites from which the experiment started. On the other hand, the cross between a white of the constitution *aabb* and any other white can never give anything but whites. For no white contains both *A* and *B*, or it would not be white, and a plant of the constitution *aabb* cannot supply the complementary factor necessary for the production of colour. Again, two whites of the constitution *Aabb* and *aaBb* when crossed should give both coloured and white flowers, the latter being three times as numerous as the former. Without going into further detail it may be stated that the results of a long series of crosses between the various F_2 whites accorded closely with the theoretical explanation.

From the evidence afforded by this exhaustive set of experiments it is impossible to resist the deduction that the appearance of colour in the sweet pea depends upon the interaction of two factors which are independently transmitted according to the ordinary scheme of Mendelian inheritance. What these factors are is still an open question. Recent evidence of a chemical nature indicates that colour in a flower is due to the interaction of two definitive substances: (1) a colourless "chromogen," or colour basis; and (2) a ferment which behaves as an activator of the chromogen, and by inducing some process of oxidation, leads to the formation of a coloured substance. But whether these two bodies exist as such {49}in the gametes or whether in some other form we have as yet no means of deciding.

Since the elucidation of the nature of colour in the sweet pea phenomena of a similar kind have been witnessed in other plants, notably in stocks, snapdragons, and orchids. Nor is this class of phenomena confined to plants. In the course of a series of experiments upon the plumage colour of poultry, indications were obtained that different white breeds did not always owe their whiteness to the same cause. Crosses were accordingly made between the white Silky fowl and a pure white strain derived from the white Dorking. Each of these had been previously shown to behave as a simple recessive to colour. When the two were crossed only fully coloured birds resulted. From analogy with the case of the sweet pea it was anticipated that such F_1 coloured birds when bred together would produce an F_2 generation consisting of coloured and white birds in the ratio 9 : 7, and when the experiment was made this was actually shown to be the case. With the growth of knowledge it is probable that further striking parallels of this nature between the plant and animal worlds will be met with.

Before quitting the subject of these experiments attention may be drawn to the fact that the 9 : 7 ratio is in reality a 9 : 3 : 3 : 1 ratio in which the last three terms are indistinguishable owing to the special circumstances that neither factor can produce a visible effect without {50}the co-operation of the other. And we may further emphasise the fact that although the two factors thus interact upon one another they are nevertheless transmitted quite independently and in accordance with the ordinary Mendelian scheme.

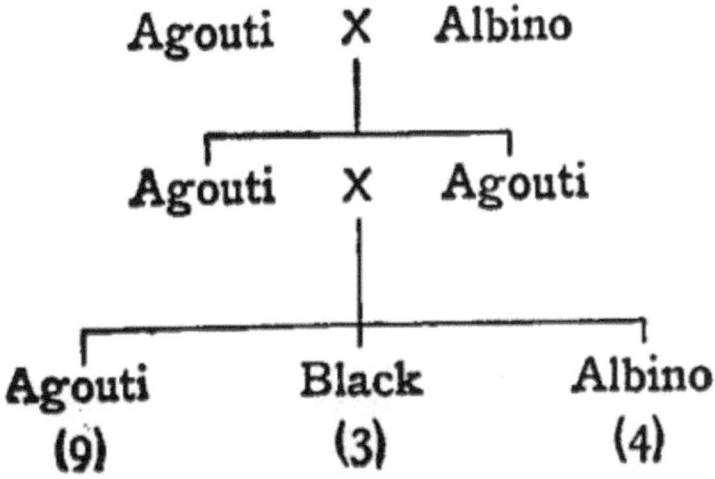

One of the earliest sets of experiments demonstrating the interaction of separate factors was that made by the French zoologist Cuénot on the coat colours of mice. It was shown that in certain cases agouti, which is the colour of the ordinary wild grey mouse, behaves as a dominant to the albino variety, *i.e.* the F_2 generation from such a cross consists of agoutis and albinos in the ratio 3 : 1. But in other cases the cross between albino and agouti gave a different result. In the F_1 generation appeared only agoutis as before, but the F_2 generation consisted of three distinct types, viz. agoutis, albinos, *and blacks*. Whence the sudden appearance of the new type? The answer is a simple one. The albino parent was really a black. But it lacked the factor without which the colour is unable to develop, and consequently it remained an albino. If we denote this factor by C, then the constitution of an albino must be cc, while that of a coloured animal may be CC or Cc, according as to whether it breeds true to colour or can {51}throw albinos. Agouti was previously known to be a simple dominant to black, *i.e.* an agouti is a black rabbit plus an additional greying factor which modifies the black into agouti. This factor we will denote by G, and we will use B for the black factor. Our original agouti and albino parents we may therefore regard as in constitution $GGCCBB$ and $ggccBB$ respectively. Both of the parents are homozygous for black. The gametes produced by the two parents are

GCB, and *gcB*, and the constitution of the F_1 animals must be *GgCcBB*. Being heterozygous for two factors they will produce four kinds of gametes in equal numbers, viz. *GCB*, *GcB*, *gCB*, and *gcB*. The results of the mating of two such similar series of gametes when the F_1 animals are bred together we may determine by the usual "chessboard" method (Fig. 8). Out of the 16 squares 9 contain both *C* and *G* in addition to *B*. Such animals must be agoutis. Three squares contain *C* but not *G*. Such animals must be coloured, but as they do not contain the modifying agouti factor their colour will be black. The remaining four squares do not contain *C*, and in the absence of this colour-developing factor they must all be albinos. Theory demands that the three classes agouti, black, and albino should appear in F_2 in the ratio 9 : 3 : 4; experiment has shown that these are the only classes that appear, and that the proportions in which they are produced accord closely with the theoretical expectation. Put briefly, then, the explanation {52}

GCB GCB Agouti	GCB GcB Agouti	GCB gCB Agouti	GCB gcB Agouti
GcB GCB Agouti	GcB GcB Albino	GcB gCB Agouti	GcB gcB Albino
gCB GCB Agouti	gCB GcB Agouti	gCB gCB BLACK	gCB gcB BLACK
gcB GCB Agouti	gcB GcB Albino	gcB gCB BLACK	gcB gcB Albino

Fig. 8.

Diagram to illustrate the nature of the F_2 generation which may arise from the mating of agouti with albino in mice or rabbits.of this case is that all the animals are black, and that we are dealing with the presence and absence of two factors, a colour developer (C), and a colour modifier (G), both acting, as it were, upon a substratum of black. The F_2 generation really consists of the four classes agoutis, blacks, albino agoutis, and albino blacks in the ratio 9 : 3 : 3 : 1. But since in the absence of the colour developer C the colour modifier G can produce no visible result, the last two classes of the ratio are indistinguishable, and our F_2 generation comes to consist of three classes in the ratio 9 : 3 : 4, instead of four classes in the ratio 9 : 3 : 3 : 1. This explanation was further tested by experiments with the albinos. In an F_2 family of this nature there ought to be three kinds, viz. albinos homozygous for G (*GGccBB*), albinos heterozygous for G (*GgccBB*), and albinos without G (*ggccBB*). These albinos are, as it were, like photographic plates exposed but undeveloped. {53}Their potentialities may be quite different, although they all look alike, but this can only be tested by treating them with a colour developer. In the case of the mice and rabbits the potentiality for which we wish to test is the presence or absence of the factor G, and in order to develop the colour we must introduce the factor C. Our developer, therefore, must contain C but not G. In other words, it must be a homozygous black mouse or rabbit, *ggCCBB*. Since such an animal is pure for C it must, when mated with any of the albinos, produce only coloured offspring. And since it does not contain G the appearance of agoutis among its offspring must be attributed to the presence of G in the albino. Tested in this way the F_2 albinos were proved, as was expected, to be of three kinds: (1) those which gave only agouti, *i.e.* which were homozygous for G; (2) those which gave agoutis and blacks in approximately equal numbers, *i.e.* which were heterozygous for G; and (3) those which gave only blacks, and therefore did not contain G.

Though albinos, whether mice, rabbits, rats, or other animals, breed true to albinism, and though albinism behaves as a simple recessive to colour, yet albinos may be of many different sorts. There are in fact just as many kinds of albinos as there are coloured forms—neither more nor less. And all these different kinds of albinos may breed together, transmitting the various colour factors according to the Mendelian scheme of inheritance, {54}and yet the visible result will be nothing but albinos. Under the mask of albinism is all the while occurring that segregation of the different colour factors which would result in all the varieties of coloured forms, if only the essential factor for colour development were present. But put in the developer by crossing with a pure coloured form and their variety of constitution can then at last become manifest.

So far we have dealt with cases in which the production of a character is dependent upon the interaction of two factors. But it may be that some characters require the simultaneous presence of a greater number of factors for their manifestation, and the experiments of Miss Saunders have shown that there is a character in stocks which is unable to appear except through the interaction of three distinct factors. Coloured stocks may be either hoary, with the leaves and stem covered by small hairs, or they may lack the hairy covering, in which case they are termed glabrous. Hoariness is dominant to glabrousness; that is to say, there is a definite factor which can turn the glabrous into a hoary plant when it is present. But in families where coloured and white stocks occur the white are always glabrous, while the coloured plants may or may not be hoary. Now colour in the stock as in the sweet pea has been proved to be dependent upon the interaction of two separate factors. Hence hoariness depends upon three separate factors, and a stock cannot be hoary unless {55}it contains the hoary factor in addition to the two colour factors. It requires the presence of all these three factors to produce the hoary character, though how this comes about we have not at present the least idea.

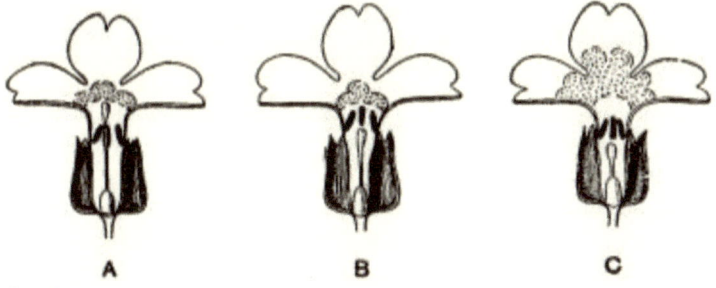

Fig. 9.

> Sections of primula flowers. The anthers are shown as black. A, "pin" form with long style and anthers set low down; B, "thrum" form with short style and anthers set higher up; C, homostyle form with anthers set low down as in "pin," but with short style. This form only occurs with the large eye.

Fig. 10.

Two primula flowers showing the extent of the small and of the large eye.

A somewhat different and less usual form of interaction between factors may be illustrated by a case in primulas recently worked out by Bateson and Gregory. Like the common primrose, the primula exhibits both pin-eyed and thrum-eyed varieties. In the former the style is long, and the centre of the eye is formed by the end of the stigma which more or less plugs up the opening of the corolla (cf. Fig. 9, A); in the latter the style is short and hidden by the four anthers which spring from higher up in the corolla and form the centre of the eye (cf. Fig. 9, B). The greater part of the "eye" is formed by the greenish-yellow patches on each petal just at the opening {56}of the corolla. In most primulas the eye is small, but there are some in which it is large and extends as a flush over a considerable part of the petals (Fig. 10). Experiments showed that these two pairs of characters behave in simple Mendelian fashion, short style (= "thrum") being dominant to long style (= "pin") and small eye dominant to large. Besides the normal long and short styled forms, there occurs a third form, which has been termed homostyle. In this form the anthers are placed low down in the corolla tube as they are in the long-styled form, but the style remains short instead of reaching up to the corolla opening (Fig. 9, C). In the course of their experiments Bateson and Gregory crossed a large-eyed homostyle plant with a small-eyed thrum (= short style). The F_1 plants were all short

styled with small eyes. {57}

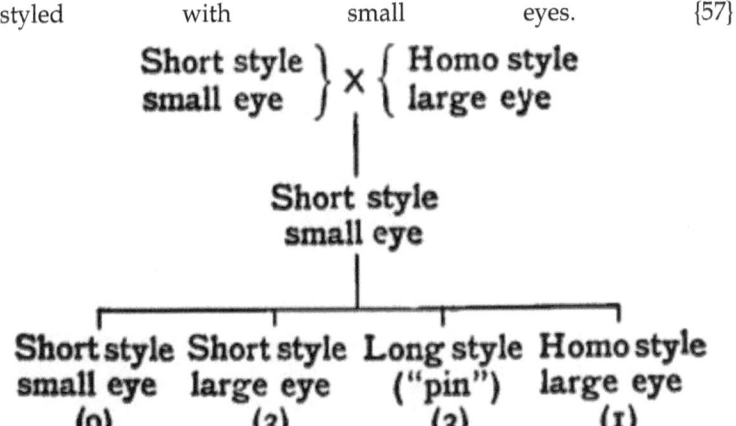

On self-fertilisation these gave an F_2 generation consisting of four types, viz. short styled with small eyes, short styled with large eyes, *long styled* with small eyes, and *homostyled* with large eyes. The notable feature of this generation is the appearance of long-styled plants, which, however, occur only in association with the small eye. The proportions in which these four types appeared shows that the presence or absence of but two factors is concerned, and at the same time provides the key to the nature of the homostyled plants. These are potentially long styled, and the position of the anthers is that of normal long-styled plants, but owing to some interaction between the factors the style itself is unable to reach its full development unless the factor for the small eye is present. For this reason long-styled plants with the large eye are always of the homostyle form. What the connecting-link between these apparently unrelated structures may be we cannot yet picture to ourselves, any more than we can picture the relation between flower {58}colour and hairiness in stocks. It is evident, however, that the conception of the interaction of factors, besides clearing up much that is paradoxical in heredity, promises to indicate lines of research which may lead to valuable extensions in our knowledge of the way in which the various parts of the living organism are related to one another.

{59}

CHAPTER VI

REVERSION

As soon as the idea was grasped that characters in plants and animals might be due to the interaction of complementary factors, it became evident that this threw clear light upon the hitherto puzzling phenomenon of reversion. We have already seen that in certain cases the cross between a black mouse or rabbit and an albino, each belonging to true breeding strains, might produce nothing but agoutis. In other words, the cross between the black and the white in certain instances results in a complete reversion to the wild grey form. Expressed in Mendelian terms, the production of the agouti was the necessary consequence of the meeting of the factors C and G in the same zygote. As soon as they are brought together, no matter in what way, the reversion is bound to occur. Reversion, therefore, in such cases we may regard as the bringing together of complementary factors which had somehow in the course of evolution become separated from one another. In the simplest cases, such as that of the black and the white rabbit, only two factors are concerned, and one of them is brought in from each of the {60}two parents. But in other cases the nature of the reversion may be more complicated owing to a larger number of factors being concerned, though the general principle remains the same. Careful breeding from the reversions will enable us in each case to determine the number and nature of the factors concerned, and in illustration of this we may take another example from rabbits. The Himalayan rabbit is a well-known breed. In appearance it is a white rabbit with pink eyes, but the ears, paws, and nose are black (Pl. I., 2). The Dutch rabbit is another well-known breed. Generally speaking, the anterior portion of the body is white, and the posterior part coloured. Anteriorly, however, the eyes are surrounded by coloured patches extending up to the ears, which are entirely coloured. At the same time the hind paws are white (cf. Pl. I., 1). Dutch rabbits exist in many varieties of colour, though in each one of these the distribution of colour and white shows the same relations. In the experiments about to be described a yellow Dutch rabbit was crossed with a Himalaya. The result was a reversion to the wild agouti colour (Pl. I., 3). Some of the F_1 individuals showed white patches, while others were self-coloured. On breeding from the F_1 animals a series of coloured forms resulted in F_2. These were agoutis, blacks, yellows, and sooty yellows, the so-called tortoise shells of the fancy (Pl. I., 4-7).

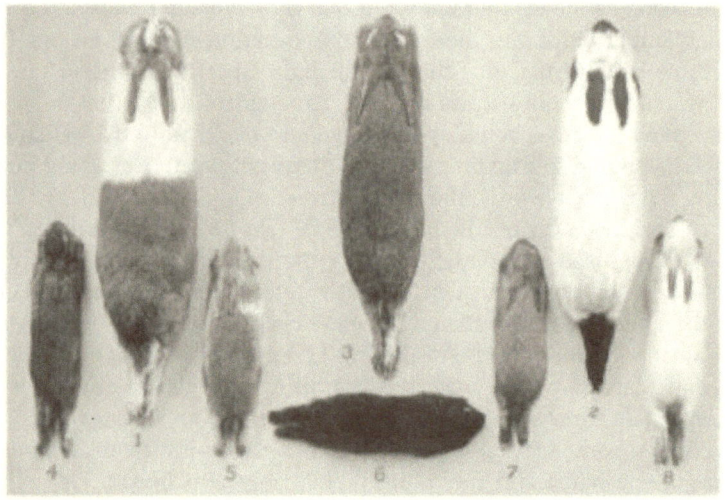

Plate I.

1, Yellow Dutch Rabbit; 2, Himalayan; 3, Agouti (= grey) F_1 reversion; 4-8, F_2 types, viz.: 4, Agouti; 5, Yellow; 6, Black; 7, Tortoise-shell; 8, Himalayan.

{61}

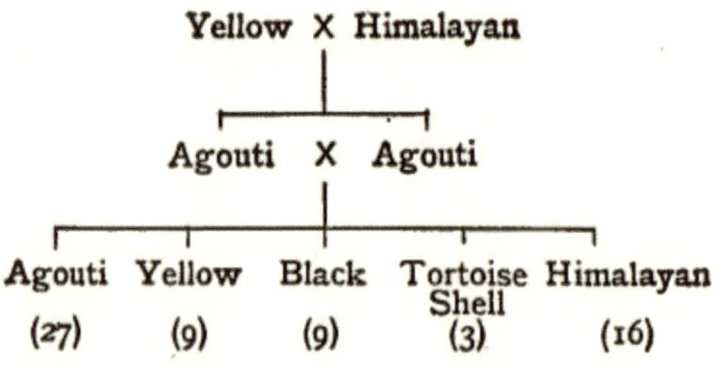

In addition to these appeared Himalayans with either black points or with lighter brownish ones, and the proportions in which they came showed the Himalayan character to be a simple recessive. A certain number of the coloured forms exhibited the Dutch marking to a greater or less extent, but as its inheritance in this set of experiments is complicated and has not yet been worked out, we may for the present neglect it and confine our attention to the coloured types and to the Himalayans. The proportion in which the four coloured types appeared in F_2 was very nearly 9 agoutis, 3 blacks, 3 yellows, and 1 tortoiseshell. Evidently we are here dealing with two factors: (1) the grey factor (G), which modifies black into agouti, or tortoiseshell into yellow; and (2) an intensifying factor (*I*), which intensifies yellow into agouti and tortoiseshell into black. It may be mentioned here that other experiments confirmed the view that the yellow rabbit is a dilute agouti, and the tortoiseshell a dilute black. The Himalayan pattern behaves as a recessive to self-colour. It is a self-coloured black rabbit lacking a factor that allows the colour to develop except in the points. That factor we may denote {62}by *X*, and as far as it is concerned the Himalayan is constitutionally *xx*. The Himalayan contains the intensifying factor, for such pigment as it possesses in the points is full coloured. At the same time it is black, *i.e.* lacking in the factor *G*. With regard to these three factors, therefore, the constitution of the Himalayan is *ggIIxx*. The last character which we have to consider in this cross is the Dutch character. This was found by Hurst to behave as a recessive to self-colour (*S*), and for our present purpose we will regard it as differing from a self-coloured rabbit in the lack of this factor. [3] The Himalayan is really a self-coloured animal, which, however, is unable to show itself as a full black owing to its not possessing the factor *X*. The results of breeding experiments then suggest that we may denote the Himalayan by the formula *ggIIxxSS* and the yellow Dutch by *GGiiXXss*. Each lacks two of the factors upon the full complement of which the agouti colour depends. By crossing them the complete series *GIXS* is brought into the same zygote, and the result is a reversion to the colour of the wild rabbit.

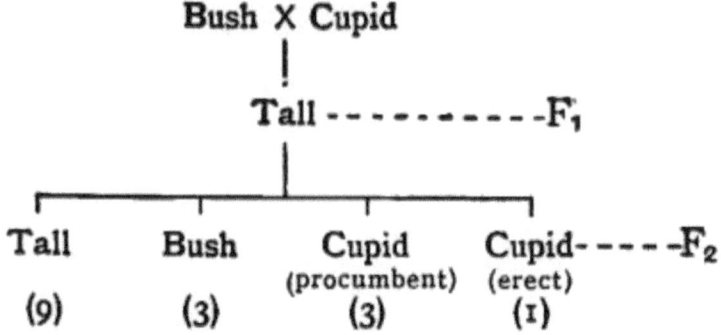

Most of the instances of reversion yet worked out are those in which colour characters are concerned. The sweet pea, however, supplies us with a good example of reversion in structural characters. A dwarf variety known as the "Cupid" has been extensively grown for {63}some years. In these little plants the internodes are very short and the stems are few in number, and attain to a length of only 9-10 inches. In course of growth they diverge from one another, and come to lie prostrate on the ground (Pl. II., 2). Curiously enough, although the whole plant is dwarfed in other respects, this does not seem to affect the size of the flower, which is that of a normal sweet pea. Another though less-known variety is the "Bush" sweet pea. Its name is derived from its habit of growth. The numerous stems do not diverge from one another, but all grow up side by side, giving the plant the appearance of a compact bush (Pl. II., 1). Under ordinary conditions it attains a height of 3½-4 feet. A number of crosses were made between the Bush and Cupid varieties, with the somewhat unexpected result that in every instance the F_1 plants showed complete reversion to the size and habit of the ordinary tall sweet pea (Pl. II., 3), which is the form of the wild plant as it occurs in Sicily to-day. The F_2 generation from these reversionary talls consisted of four different types, viz. {64}talls, bushes, Cupids of the procumbent type like the original Cupid parent, and Cupids with the compact upright Bush habit (Pl. II., 4). These four types appeared in the ratio $9:3:3:1$, and this, of course, provided the clue to the nature of the case. The characters concerned are (1) long internode of stem between the leaves which is dominant to short internode, and (2) the creeping procumbent habit which is dominant to the erect bush-like habit. Of these characters length of internode was carried by the Bush, and the procumbent habit by the original Cupid parent. The bringing of them together by the cross resulted in a procumbent plant with long internodes. This is the ordinary tall sweet pea of the wild Sicilian type, reversion here, again, being due to the bringing together of two complementary factors which had somehow become separated in the course of evolution.

To this interpretation it may be objected that the ordinary sweet pea is a plant of upright habit. This, however, is not true. It only appears so because the conventional way of growing it is to train it up sticks. In reality it is of procumbent habit, with divergent stems like the ordinary Cupid, a fact which can easily be observed by anyone who will watch them grow without the artificial aid of prepared supports.

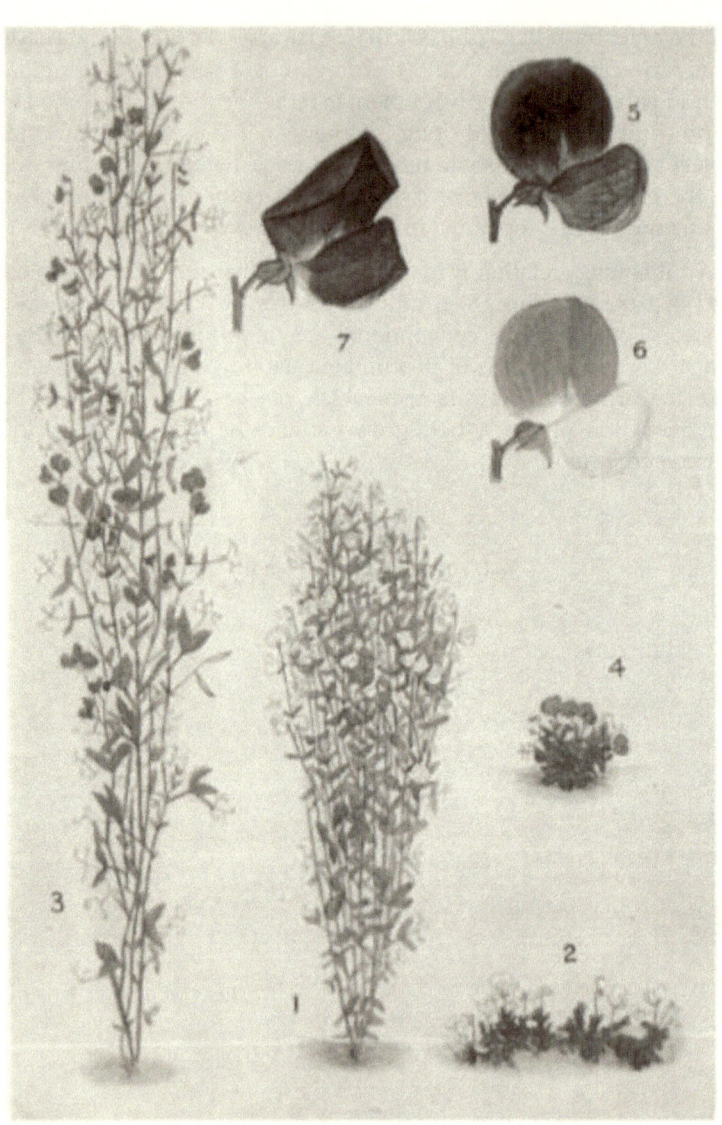

Plate II.

1, Bush Sweet Pea; 2, Cupid Sweet Pea; 3, F_1 reversionary Tall; 4, Erect Cupid Sweet Pea; 5, Purple Invincible; 6, Painted Lady; 7, Duke of Westminster (hooded standard).

{65}

The cases of reversion with which we have so far dealt have been cases in which the reversion occurs as an immediate result of a cross, *i.e.* in the F_1 generation. This is perhaps the commonest mode of reversion, but instances are known in which the reversion that occurs when two pure types are crossed does not appear until the F_2 generation. Such a case we have already met with in the fowls' combs. It will be remembered that the cross between pure pea and pure rose gave walnut combs in F_1, while in the F_2 generation a definite proportion, 1 in 16, of single combs appeared (cf. p. 32). Now the single comb is the form that is found in the wild jungle fowl, which is generally regarded as the ancestor of the domestic breeds. If this is so, we have a case of reversion in F_2; and this in the *absence* of the two factors brought together by the rose-comb and pea-comb parents. Instead of the reversion being due to the bringing together of two complementary factors, we must regard it here as due to the association of two complementary absences. To this question, however, we shall revert later in discussing the origin of domesticated varieties.

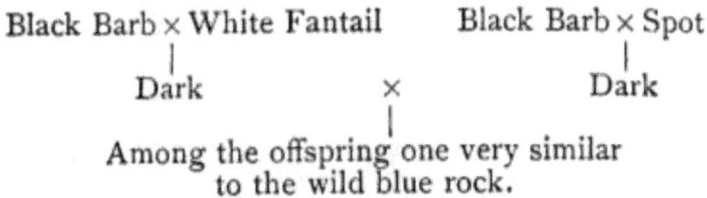

There is one other instance of reversion to which we must allude. This is Darwin's famous case of the occasional appearance of pigeons reverting to the wild blue rock (*Columba livia*), when certain domesticated races are crossed together. [4] As is well known, Darwin made use of this as an argument for regarding all the domesticated varieties as having arisen from the same wild species. The original experiment is somewhat complicated, and is shown in the accompanying scheme. Essentially it lay in {66}following the results flowing from crosses between blacks and whites. Experiments recently made by Staples-Browne have shown that this case of reversion also can be readily interpreted in Mendelian terms. In these experiments the cross was made between black barbs and white fantails.

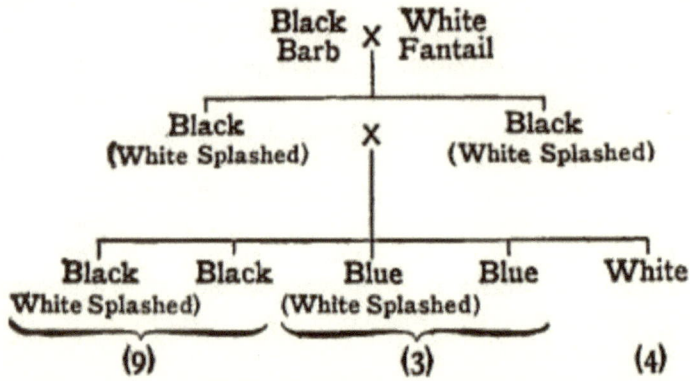

The F_1 birds were all black with some white splashes, evidently due to a separate factor introduced by the fantail. On breeding these blacks together they gave an F_2 generation, consisting of blacks (with or without white splashes), blues (with or without white splashes), and whites in the ratio 9 : 3 : 4. The factors concerned are colour (*C*), in the absence of {67}which a bird is white, and a black modifier (*B*), in the absence of which a coloured bird is blue. The original black barb contained both of these factors, being in constitution *CCBB*. The fantail, however, contained neither, and was constitutionally *ccbb*. The F_1 birds produced by crossing were in constitution *CcBb*, and being heterozygous for two factors produced in equal numbers the four sorts of gametes *CB*, *Cb*, *cB*, *cb*. The results of two such series of gametes being brought together are shown in the usual way in Fig. 11. A blue is a bird containing the colour factor but lacking the black modifier, *i.e.* of the constitution *CCbb*, or *Ccbb*, and such birds as the figure shows appear in the F_2 generation on the average three times out of sixteen. Reversion here comes about in F_2, when the redistribution of the factors leads to the formation of zygotes containing one of the two factors but not the other.

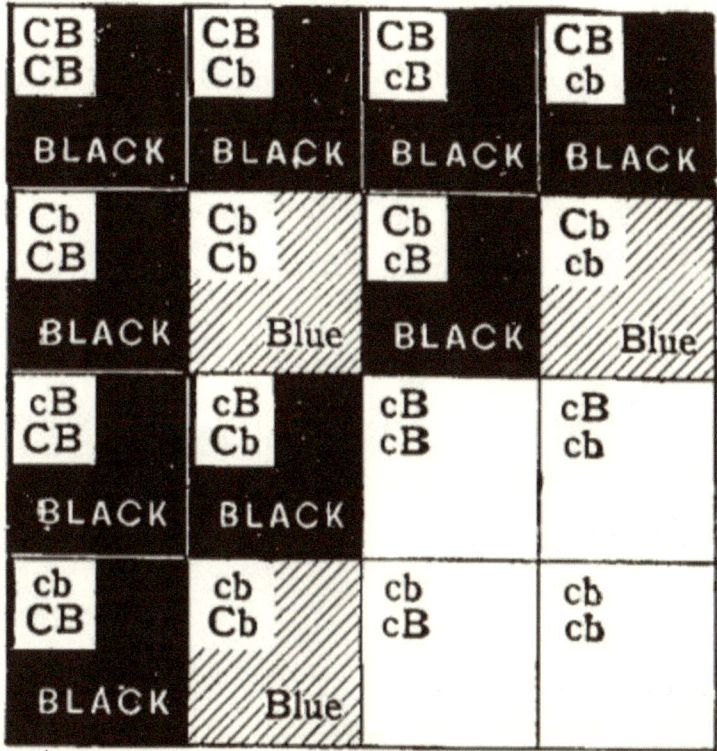

Fig. 11.

Diagram to illustrate the appearance of the reversionary blue pigeon in F₂ from the cross of black with white.

{68}

CHAPTER VII

DOMINANCE

Fig. 12.

Primula flowers to illustrate the intermediate nature of the F_1 flower when *sinensis* is crossed with *stellata*.

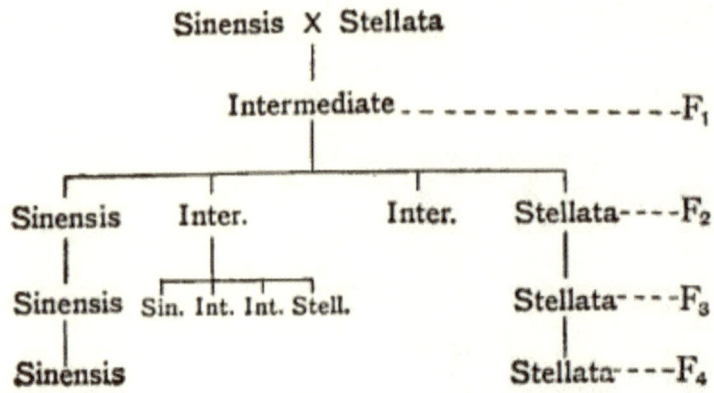

In the cases which we have hitherto considered the presence of a factor produces its full effect whether it is introduced by both of the gametes which go to form the zygote, or by one of them alone. The heterozygous tall pea or the heterozygous rose-combed fowl cannot be distinguished from the homozygous form by mere inspection, however close. Breeding tests alone can decide which is the heterozygous and which the homozygous form. Though this is true for the majority of characters yet investigated, there are cases known in which the heterozygous form differs in appearance from either parent. Among plants such a case has been met with in the primula. The ordinary Chinese primula (*P. sinensis*) (Fig. 12) has large rather wavy petals much crenated at the edges. In the Star Primula (*P. stellata*) the flowers are much smaller, while the petals are flat and present only a terminal notch instead of the numerous crenations of *P. sinensis*. The heterozygote produced by crossing these forms is intermediate in size and appearance. When self-fertilised such plants behave in simple Mendelian fashion, {69}giving a generation consisting of *sinensis*, *intermediates*, and *stellata* in the ratio 1 : 2 : 1. Subsequent breeding from these plants showed that both the *sinensis* and *stellata* which appeared in the F_2 generation bred true, while the intermediates always gave all three forms again in the same proportion. But though there is no dominance of the character of either parent in such a case as this, the Mendelian principle of segregation could hardly have a better illustration.

{70}

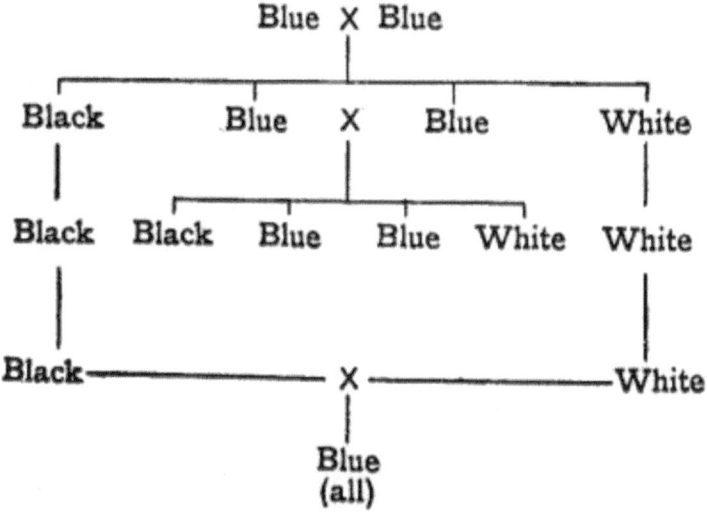

Among birds a case of similar nature is that of the Blue Andalusian fowl. Fanciers have long recognised the difficulty of getting this variety to breed true. Of a slaty blue colour itself with darker hackles and with black lacing on the feathers of the breast, it always throws "wasters" of two kinds, viz. blacks, and whites splashed with black. Careful breeding from the blues shows that the three sorts are always produced in the same definite {71}proportions, viz., one black, two blues, one splashed white. This at once suggests that the black and the splashed white are the two homozygous forms, and that the blues are heterozygous, *i.e.*, producing equal numbers of "black" and "white splashed" gametes. The view was tested by breeding the "wasters" together—black with black, and splashed white with splashed white—and it was found that each bred true to its respective type. But when the black and the splashed white were crossed they gave, as was expected, nothing but blues. In other words, we have the seeming paradox of the black and the splashed white producing twice as many blues as do the blues when bred together. The black and the splashed white "wasters" are in reality the pure breeds, while the "pure" Blue Andalusian is a mongrel which no amount of selection will ever be able to fix.

In such cases as this it is obvious that we cannot speak of dominance. And with the disappearance of this phenomenon we lose one criterion for determining which of the two parent forms possesses the additional factor. Are we, for example, to regard the black Andalusian as a splashed white to which has been added a double dose of a colour-intensifying factor, or are we to consider the white splashed bird as a black which is unable to show its true pigmentation owing to the possession of some inhibiting factor which prevents the manifestation of the black. Either interpretation fits the facts equally well, {72}and until further experiments have been devised and carried out it is not possible to decide which is the correct view.

Besides these comparatively rare cases where the heterozygote cannot be said to bear a closer resemblance to one parent more than to the other, there are cases in which it is often possible to draw a visible distinction between the heterozygote and the pure dominant. There are certain white breeds of poultry, notably the White Leghorn, in which the white behaves as a dominant to colour. But the heterozygous whites made by crossing the dominant white birds with a pure coloured form (such as the Brown Leghorn) almost invariably show a few coloured feathers or "ticks" in their plumage. The dominance of white is not quite complete, and renders it possible to distinguish the pure from the impure dominant without recourse to breeding experiments.

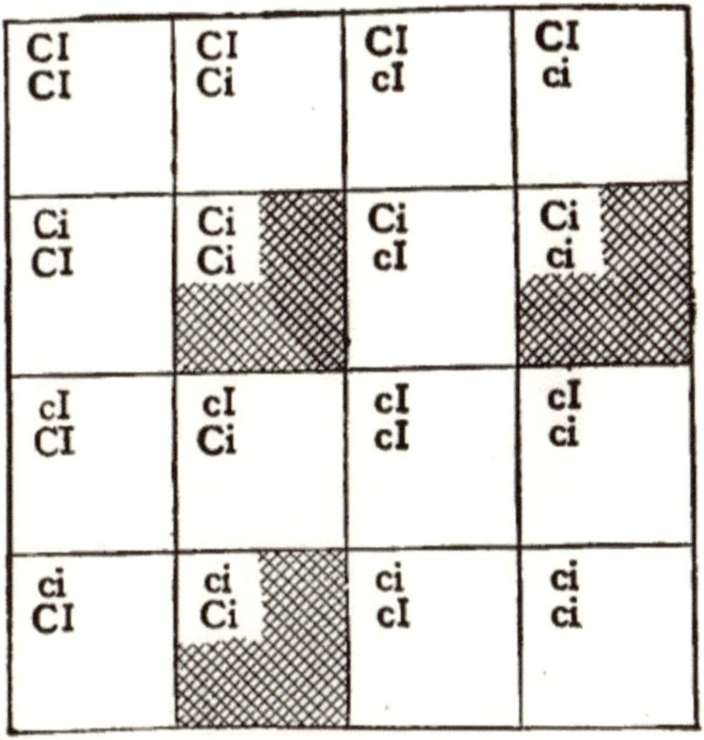

CI CI	CI Ci	CI cI	CI ci
Ci CI	Ci Ci	Ci cI	Ci ci
cI CI	cI Ci	cI cI	cI ci
ci CI	ci Ci	ci cI	ci ci

Fig. 13.

Diagram to illustrate the nature of the F_2 generation from the cross between dominant white and recessive white fowls.

This case of the dominant white fowl opens up another interesting problem in connection with dominance. By accepting the "Presence and Absence" hypothesis we are committed to the view that the dominant form possesses an extra factor as compared with the recessive. The natural way of looking at this case of the fowl is to regard white as the absence of colour. But were this so, colour should be dominant to white, which is not the case. We are therefore forced to suppose that the absence of colour in this instance is due to the presence of a factor whose {73}property is to inhibit the production of colour in what would otherwise be a pure coloured bird. On this view the dominant white fowl is a coloured bird plus a factor which inhibits the development of the colour. The view can be put to the test of experiment. We have already seen that there are other white fowls in which white is recessive to colour, and that the whiteness of such birds is due to the fact that they lack a factor for the development of colour. If we denote this factor by C and our postulated inhibitor factor in the dominant white bird by I, then we must write the constitution of the recessive white as *ccii*, and the dominant white as *CCII*. We may now work out the results we ought to obtain when a cross is made between these two pure white breeds. The constitution of the F_1 bird must be *CcIi*. Such birds being heterozygous for the inhibitor factor, should be whites showing some coloured "ticks." Being heterozygous for both of the two factors C and I, they will produce in equal numbers the four different sorts of gametes *CI*, *Ci*, *cI*, *ci*. The result of bringing two such similar series of gametes together is shown in Fig. 13. Out of the sixteen squares, twelve contain *I*; these will be white birds either with or without a few coloured ticks. Three contain C but not *I*: these must be coloured birds. One contains neither C nor *I*; this must be a white. From such a mating we ought, therefore, to obtain both white and coloured birds in the ratio 13 : 3. The results thus theoretically {74}deduced were found to accord with the actual facts of experiment. The F_1 birds were all "ticked" whites, and in the F_2 generation came white and coloured birds in the expected ratio. There seems, therefore, little reason to doubt that the dominant white is a coloured bird in which the absence of colour is due to the action of a colour-inhibiting factor, though as to the nature of that factor we can at present make no surmise. It is probable that other facts, which at first sight do not appear to be in agreement with the "Presence and Absence" hypothesis, will eventually be brought into line through the action of inhibitor factors. Such a case, for instance, is that of bearded and beardless wheats. Though the beard is obviously the additional character, the bearded condition is recessive to the beardless. Probably we ought to regard the beardless as a bearded wheat in which there is an inhibitor that stops the beard from growing. It is not unlikely that as time goes on we shall {75}find many more such cases of the action of inhibitor factors, and we must be prepared to find that the same visible effect may be produced either by the addition or by the omission of a factor. The dominant and recessive white poultry are indistinguishable in appearance. Yet the one contains a factor more and the other a factor less than the coloured bird.

Fig. 14.

Ears of beardless and bearded wheat. The beardless condition is dominant to the bearded.

A phenomenon sometimes termed irregularity of dominance has been investigated in a few cases. In certain breeds of poultry such as Dorkings there occurs an extra toe directed backwards like the hallux (cf. Fig. 15). In some families this character behaves as an ordinary dominant to the normal, giving the expected 3 : 1 ratio in F_2. But in other families similarly bred the proportions of birds with and without the extra toe appear to be unusual. It has been shown that in such a family some of the birds without the extra toe may nevertheless transmit the peculiarity when mated with birds belonging to strains in which the extra toe never occurs. Though the external appearance of the bird generally affords some indication of the nature of the gametes which it is carrying, this is not always the case. Nevertheless we have reason to suppose that the character segregates in the gametes, though the nature of these cannot always be decided from the appearance of the bird which bears them.

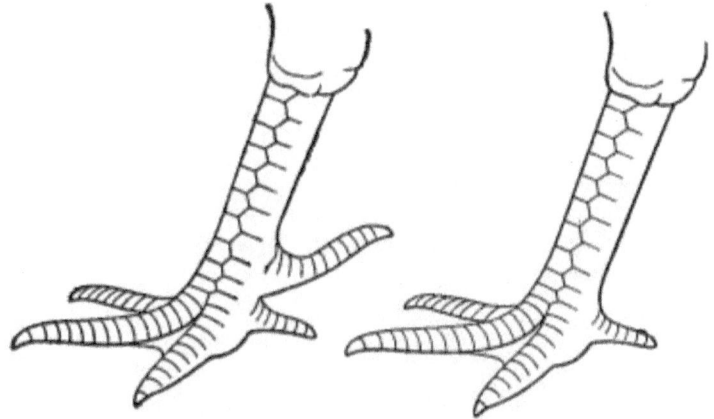

Fig. 15.

Fowls' feet. On the right a normal and on the left one with an extra toe.

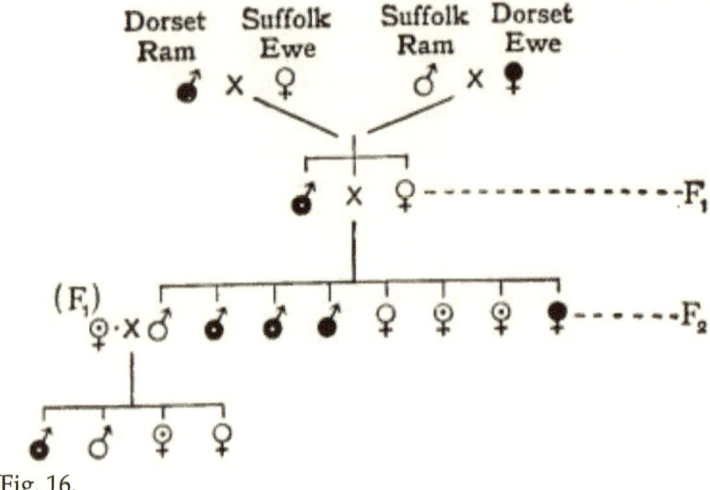

Fig. 16.

Scheme to illustrate the inheritance of horns in sheep. Heterozygous males shown dark with a white spot, heterozygous females light with a dark spot in the centre.

There are cases in which an apparent irregularity of dominance has been shown to depend upon another character, as in the experiments with sheep carried out by Professor Wood. In these experiments two breeds were crossed, of which one, the Dorset, is horned in both sexes, while the other, the Suffolk, is without horns in either sex. Whichever way the cross was made the resulting F_1 generation was similar; the rams were horned, and {77}the ewes were hornless. In the F_2 generation raised from these F_1 animals both horned and hornless types appeared in both sexes but in very different proportions. While the horned rams were about three times as numerous as the hornless, this relation was reversed among the females, in which the horned formed only about one-quarter of the total. The simplest explanation of this interesting case is to suppose that the dominance of the horned character depends upon the sex of the animal—that it is dominant in the male but recessive in the female. A pretty experiment was devised for putting this view to the test. If it is true, equal numbers of gametes with and without the horned factor must be produced by the F_1 ewes, while the factor should be lacking in all the gametes of the hornless F_2 rams. A {78}hornless ram, therefore, put to a flock of F_1 ewes should give rise to equal numbers of zygotes which are heterozygous for the horned character, and of zygotes in which it is completely absent. And since the heterozygous males are horned, while the heterozgyous females are hornless, we should expect from this mating equal numbers of horned and hornless rams, but only hornless ewes. The result of the experiment confirmed this expectation. Of the ram lambs 9 were horned and 8 were hornless, while all the 11 ewe lambs were completely destitute of horns.

Plate III.
Sheep
{79}

CHAPTER VIII

WILD FORMS AND DOMESTIC VARIETIES

In discussing the phenomena of reversion we have seen that in most cases such reversion occurs when the two varieties which are crossed each contain certain factors lacking in the other, of which the full complement is necessary for the production of the reversionary wild form. This at once suggests the idea that the various domestic forms of animals and plants have arisen by the omission from time to time of this factor or of that. In some cases we have clear evidence that this is the most natural interpretation of the relation between the cultivated and the wild forms. Probably the species in which it is most evident is the sweet pea (*Lathyrus odoratus*). We have already seen reason to suppose that as regards certain structural features the Bush variety is a wild lacking the factor for the procumbent habit, that the Cupid is a wild without the factor for the long inter-node, and that the Bush Cupid is a wild minus both these factors. Nor is the evidence less clear for the many colour varieties. In illustration we may consider in more detail a case in which the cross between two whites resulted {80}in a complete reversion to the purple colour characteristic of the wild Sicilian form (Pl. IV.). In this particular instance subsequent breeding from the purples resulted in the production of six different colour forms in addition to whites. The proportion of the coloured forms to the whites was 9 : 7 (cf. p. 44), but it is with the relation of the six coloured forms that we are concerned here. Of these six forms three were purples and three were reds. The three purple forms were (1) the wild bicolor purple with blue wings known in cultivation as the Purple Invincible (Pl. IV., 4); (2) a deep purple with purple wings (Pl. IV., 5); and (3) a very dilute purple known as the Picotee (Pl. IV., 6). Corresponding to these three purple forms were three reds: (1) a bicolor red known as Painted Lady (Pl. IV., 7); (2) a deep red with red wings known as Miss Hunt (Pl. IV., 8); and (3) a very pale red which we have termed Tinged White [5] (Pl. IV., 9). In the F_2 generation the total number of purples bore to the total number of reds the ratio 3 : 1, and this ratio was maintained for each of the corresponding classes. Purple, therefore, is dominant to red, and each of the three classes of red differs from its corresponding purple in not possessing the blue factor (B) which turns it into purple.

Plate IV.

1, 2, Emily Henderson; 3, F_1 reversionary Purple; 4-10, Various F_2 forms: 4, Purple; 5, Deep Purple; 6, Picotee; 7, Painted Lady; 8, Miss Hunt; 9, Tinged White; 10, White.

{81}

Again, the proportion in which the three classes of purples appeared was 9 bicolors, 3 deep purples, 4 picotees. We are, therefore, concerned here with the operation of two factors: (1) a light wing factor, which renders the bicolor dominant to the dark winged form; and (2) a factor for intense colour, which occurs in the bicolor and in the deep purple, but is lacking in the dilute picotee. And here it should be mentioned that these conclusions rest upon an exhaustive set of experiments involving the breeding of many thousands of plants. In this cross, therefore, we are concerned with the presence or absence of five factors, which we may denote as follows:—

A colour base, R.

A colour developer, C.

A purple factor, B.

A light wing factor, L.

A factor for intense colour, I.

On this notation our six coloured forms are:—

(1) Purple bicolor $CRBLI$. [6]
(2) Deep purple $CRBlI$.
(3) Picotee $CRBLi$ or $CRBli$.
(4) Red bicolor (= Painted Lady) $CRbLI$.
(5) Deep red (= Miss Hunt) $CRblI$.
(6) Tinged white $CRbLi$ or $CRbli$.

It will be noticed in this series that the various coloured {82}forms can be expressed by the omission of one or more factors from the purple bicolor of the wild type. With the complete omission of each factor a new colour type results, and it is difficult to resist the inference that the various cultivated forms of the sweet pea have arisen from the wild by some process of this kind. Such a view tallies with what we know of the behaviour of the wild form when crossed by any of the garden varieties. Wherever such crossing has been made the form of the hybrid has been that of the wild, thus supporting the view that the wild contains a complete set of all the differentiating factors which are to be found in the sweet pea.

Moreover, this view is in harmony with such historical evidence as is to be gleaned from botanical literature, and from old seedsmen's catalogues. The wild sweet pea first reached England in 1699, having been sent from Sicily by the monk Franciscus Cupani as a present to a certain Dr. Uvedale in the county of Middlesex. Somewhat later we hear of two new varieties, the red bicolor, or Painted Lady, and the white, each of which may be regarded as having "sported" from the wild purple by the omission of the purple factor, or of one of the two colour factors. In 1793 we find a seedsman offering also what he called black and scarlet varieties. It is probable that these were our deep purple and Miss Hunt varieties, and that somewhere about this time the factor for the {83}light wing (L) was dropped out in certain plants. In 1860 we have evidence that the pale purple or Picotee, and with it doubtless the Tinged White, had come into existence. This time it was the factor for intense colour which had dropped out. And so the story goes on until the present day, and it is now possible to express by the same simple method the relation of the modern shades, of purple and reds, of blues and pinks, of hooded and wavy standards, to one another and to the original wild form. The constitution of many of these has now been worked out, and to-day it would be a simple though perhaps tedious task to denote all the different varieties by a series of letters indicating the factors which they contain, instead of by the present system of calling them after kings and queens, and famous generals, and ladies more or less well known.

From what we know of the history of the various strains of sweet peas one thing stands out clearly. The new character does not arise from a pre-existing variety by any process of gradual selection, conscious or otherwise. It turns up suddenly complete in itself, and thereafter it can be associated by crossing with other existing characters to produce a gamut of new varieties. If, for example, the character of hooding in the standard (cf. Pl. II., 7) suddenly turned up in such a family as that shown on Plate IV. we should be able to get a hooded form corresponding to each of the forms with the erect {84}standard; in other words, the arrival of the new form would give us the possibility of fourteen varieties instead of seven. As we know, the hooded character already exists. It is recessive to the erect standard, and we have reason to suppose that it arose as a sudden sport by the omission of the factor in whose presence the standard assumes the erect shape characteristic of the wild flower. It is largely by keeping his eyes open and seizing upon such sports for crossing purposes that the horticulturist "improves" the plants with which he deals. How these sports or *mutations* come about we can now surmise. They must owe their origin to a disturbance in the processes of cell division through which the gametes originate. At some stage or other the normal equal distribution of the various factors is upset, and some of the gametes receive a factor less than others. From the union of two such gametes, provided that they are still capable of fertilisation, comes the zygote which in course of growth develops the new character.

Why these mutations arise: what leads to the surmised unequal division of the gametes: of this we know practically nothing. Nor until we can induce the production of mutations at will are we likely to understand the conditions which govern their formation. Nevertheless there are already hints scattered about the recent literature of experimental biology which lead us to hope that we may know more of these matters in the future. {85}

In respect of the evolution of its now multitudinous varieties, the story of the sweet pea is clear and straightforward. These have all arisen from the wild by a process of continuous loss. Everything was there in the beginning, and as the wild plant parted with factor after factor there came into being the long series of derived forms. Exquisite as are the results of civilization, it is by the degradation of the wild that they have been brought about. How far are we justified in regarding this as a picture of the manner in which evolution works?

There are certainly other species in which we must suppose that this is the way that the various domesticated forms have arisen. Such, for example, is the case in the rabbit, where most of the colour varieties are recessive to the wild agouti form. Such also is the case in the rat, where the black and albino varieties and the various pattern forms are also recessive to the wild agouti type. And with the exception of a certain yellow variety to which we shall refer later, such is also the case with the many fancy varieties of mice.

Nevertheless there are other cases in which we must suppose evolution to have proceeded by the interpolation of characters. In discussing reversion on crossing, we have already seen that this may not occur until the F_2 generation, as, for example, in the instance of the fowls' combs (cf. p. 65). The reversion to the single comb occurred as the result of the removal of the two factors {86}for rose and pea. These two domesticated varieties must be regarded as each possessing an additional factor in comparison with the wild single-combed bird. During the evolution of the fowl, these two factors must be conceived of as having been interpolated in some way. And the same holds good for the inhibitory factor on which, as we have seen, the dominant white character of certain poultry depends. In pigeons, too, if we regard the blue rock as the ancestor of the domesticated breeds, we must suppose that an additional melanic factor has arisen at some stage. For we have already seen that black is dominant to blue, and the characters of F_1, together with the greater number of blacks than blues in F_2, negatives the possibility that we are here dealing with an inhibitory factor. The hornless or polled condition of cattle, again, is dominant to the horned condition, and if, as seems reasonable, we regard the original ancestors of domestic cattle as having been horned, we have here again the interpolation of an inhibitory factor somewhere in the course of evolution.

On the whole, therefore, we must be prepared to admit that the evolution of domestic varieties may come about by a process of addition of factors in some cases and of subtraction in others. It may be that what we term additional factors fall into distinct categories from the rest. So far, experiment seems to show that they are either of the nature of melanic factors, or of inhibitory {87}factors, or of reduplication factors as in the case of the fowls' combs. But while the data remain so scanty, speculation in these matters is too hazardous to be profitable.

CHAPTER IX

REPULSION AND COUPLING OF FACTORS

Although different factors may act together to produce specific results in the zygote through their interaction, yet in all the cases we have hitherto considered the heredity of each of the different factors is entirely independent. The interaction of the factors affects the characters of the zygote, but makes no difference to the distribution of the separate factors, which is always in strict accordance with the ordinary Mendelian scheme. Each factor in this respect behaves as though the other were not present.

A few cases have been worked out in which the distribution of the different factors to the gametes is affected by their simultaneous presence in the zygote. And the influence which they are able to exert upon one another in such cases is of two kinds. They may repel one another, refusing, as it were, to enter into the same zygote, or they may attract one another, and, becoming linked together, pass into the same gamete, as it were by preference. For the moment we may consider these two sets of phenomena apart. {89}

One of the best illustrations of repulsion between factors occurs in the sweet pea. We have already seen that the loss of the blue or purple factor (B) from the wild bicolor results in the formation of the red bicolor known as Painted Lady (Pl. IV., 7). Further, we have seen that the hooded standard is recessive to the ordinary erect standard. The omission of the factor for the erect standard (E) from the purple bicolor (Pl. II., 5) results in a hooded purple known as Duke of Westminster (Pl. II., 7). And here it should be mentioned that in the corresponding hooded forms the difference in colour between the wings and standard is not nearly so marked as in the forms with the erect standard, but the difference in structure appears to affect the colour, which becomes nearly uniform. This may be readily seen by comparing the picture of the purple bicolor on Plate II. with that of the Duke of Westminster flower.

Now when a Duke of Westminster is mated with a Painted Lady the factor for erect standard (E) is brought in by the red, and that for blue (B) by the Duke, and the offspring are consequently all purple

89

bicolors. Purples so formed are all heterozygous for these two factors, and were the case a simple one, such as those which have already been discussed, we should expect the F₂ generation to consist of the four forms: erect purple, hooded purple, erect red, and hooded red in the ratio 9 : 3 : 3 : 1. Such, however, is not the case. The F₂ generation {90}actually consists of only three forms, viz. erect red, erect purple, and hooded purple, and the ratio in which these three forms occur is 1 : 2 : 1. No hooded red has been known to occur in such a family. Moreover further breeding shows that while the erect reds and the hooded purples always breed true, the erect purples in such families *never* breed true, but always behave like the original F₁ plant, giving the three forms again in the ratio 1 : 2 : 1. Yet we know that there is no difficulty in getting purple bicolors to breed true from other families; and we know also that hooded red sweet peas exist in other strains.

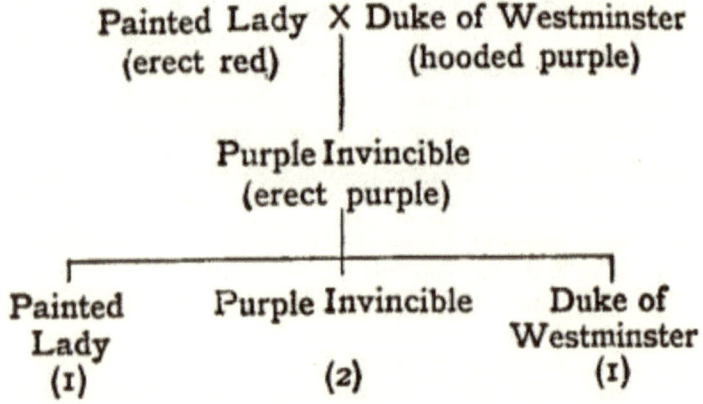

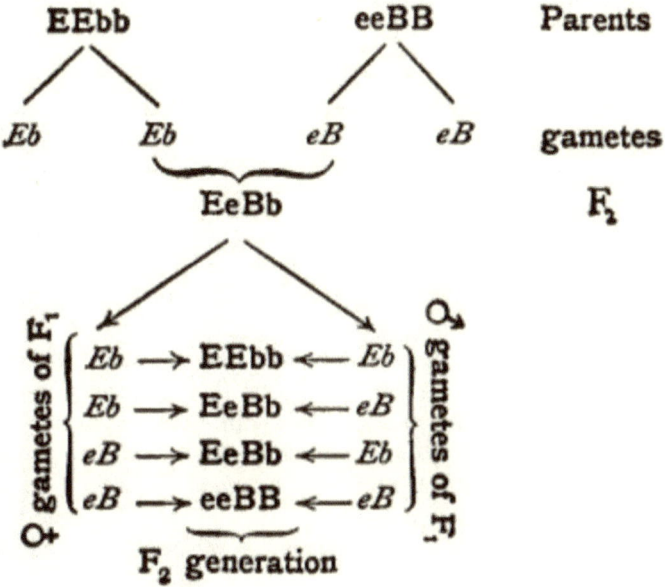

On the assumption that there exists a repulsion between the factors for erect standard and blue in a plant which is heterozygous for both, this peculiar case receives a simple explanation. The constitutions of the erect red and the hooded purple are *EEbb* and *eeBB* respectively and that of the F₁ erect purple is *EeBb*. Now let us suppose that in such a zygote there exists a repulsion {91}between *E* and *B*, such that when the plant forms gametes these two factors will not go into the same gamete. On this view it can only form two kinds of gametes, viz. *Eb* and *eB*, and these, of course, will be formed in equal numbers. Such a plant on self-fertilisation must give the zygotic series *EEbb* + 2 *EeBb* + *eeBB*, *i.e.* 1 erect red, 2 erect purples, and 1 hooded purple. And because the erect reds and the hooded purples are respectively homozygous for *E* and *B*, they must thenceforward breed true. The erect purples, on the other hand, being always formed by the union of a gamete *Eb* with a gamete *eB*, are always heterozygous for both of these factors. They can, consequently, never breed true, but must always give erect reds, erect purples, and hooded purples in the ratio 1 : 2 : 1. The experimental facts are readily explained on the assumption of repulsion between the two {92}factors B and E during the formation of the gametes in a plant which is heterozygous for both.

Other similar cases of factorial repulsion have been demonstrated in the sweet pea, and two of these are also concerned with the two factors with which we have just been dealing. Two distinct varieties of pollen grains occur in this species, viz. the ordinary oblong form and a rather smaller rounded grain. The former is dominant to the latter. [7] When a cross is made between a purple with round pollen and a red with long pollen the F_1 plant is a long pollened purple. But the F_2 generation consists of purples with round pollen, purples with long pollen, and reds with long pollen in the ratio 1 : 2 : 1. No red with round pollen appears in F_2 owing to repulsion between the factors for purple (B) and for long pollen (L). Similarly plants produced by crossing a red hooded long with a red round having an erect standard give in F_1 long pollened reds with an erect standard, and these in F_2 produce the three types, round pollened erect, long pollened erect, and long pollened hooded, in the ratio 1 : 2 : 1. The repulsion here is between the long pollen factor (L) and the factor for the erect standard (E).

{93}

Yet another similar case is known in which we are concerned with quite different factors. In some sweet peas the axils whence the leaves and flower stalks spring from the main stem are of a deep red colour. In others they are green. The dark pigmented axil is dominant to the light one. Again, in some sweet peas the anthers are sterile, setting no pollen, and this condition is recessive to the ordinary fertile condition. When a sterile plant with a dark axil is crossed by a fertile plant with a light axil, the F_1 plants are all fertile with dark axils. But such plants in F_2 give fertiles with light axils, fertiles with dark axils, and steriles with dark axils in the ratio 1 : 2 : 1. No light axilled steriles appear from such a cross owing to the repulsion between the factor for dark axil (D) and that for the fertile anther (F).

These four cases have already been found in the sweet pea, and similar phenomena have been met with by Gregory in primulas. To certain seemingly analogous cases in animals where sex is concerned we shall refer later.

Now all of these four cases present a common feature which probably has not escaped the attention of the reader. In all of them *the original cross was such as to introduce one of the repelling factors with each of the two parents*. If we denote our two factors by *A* and *B*, the crosses have always been of the nature *AAbb* × *aaBB*. Let us now consider what happens when both of the {94}factors, which in these cases repel one another, are introduced by one of the parents, and neither by the other parent. And in particular we will take the case in which we are concerned with purple and red flower colour, and with long and round pollen, *i.e.* with the factors *B* and *L*. When a purple long (*BBLL*) is crossed with a red round (*bbll*) the F_1 (*BbLl*) is a purple with long pollen, identical in appearance with that produced by crossing the long pollened red with the round pollened purple. But the nature of the F_2 generation is in some respects very different. The ratio of purples to reds and of longs to rounds is in each case 3 : 1, as before. But instead of an association between the red and the long pollen characters the reverse is the case. The long pollen character is now associated with purple and the round pollen with red. The association, however, is not quite complete, and the examination of a large quantity of similarly bred material shows that the purple longs are about twelve times as numerous as the purple rounds, while the red rounds are rather more than three times as many as the red longs. Now this peculiar result could be brought about if the gametic series produced by the F_1 plant consisted of 7 *BL* + 1 *Bl* + 1 *bL* + 7 *bl* out of every 16 gametes. Fertilization between two such similar series of 16 gametes would result in 256 plants, of which 177 would be purple longs, 15 purple rounds, 15 red longs, and 49 red rounds—a proportion of the four different kinds very close to {95}that actually found by experiment. It will be noticed that in the whole family the purples are to the reds as 3 : 1, and the longs are also three times as numerous as the rounds. The peculiarity of the case lies in the distribution of these two characters with regard to one another. In some way or other the factors for blue and for long pollen become linked together in the cell divisions that give rise to the gametes, but the linking is not complete. This holds good for all the four cases in which repulsion between the factors occurs when one of the two factors is introduced by each of the parents. *When both of the factors are brought into the cross by the same parent we get coupling between them instead of repulsion*. The phenomena of repulsion and coupling between separate factors are intimately related, though hitherto we have not been able to suggest why this should be so.

Nor for the present can we suggest why certain factors should be linked together in the peculiar way that we have reason to suppose that they are during the process of the formation of the gametes. Nevertheless the phenomena are very definite, and it is not unlikely that a further study of them may throw important light on the architecture of the living cell.

APPENDIX TO CHAPTER IX

As it is possible that some readers may care, in spite of its complexity, to enter rather more fully into the peculiar phenomenon {96}of the coupling of characters, I have brought together some further data in this Appendix. In the case we have already considered, where the factors for blue colour and long pollen are concerned, we have been led to suppose that the gametes produced by the heterozygous plant are of the nature 7 BL : 1 Bl : 1 bL : 7 bl. Such a series of ovules fertilised by a similar series of pollen grains will give a generation of the following composition:—

49 BBLL	+ 7 BBLl	+ 7 BbLL	+ 49	BbLl	+ BBll
	+ 7 BBLl	+ 7 BbLL	+	BbLl	
			+	BbLl	
			+ 49	BbLl	

177 purple, long

and as this theoretical result fits closely with the actual figures obtained by experiment we have reason for supposing that the heterozygous plant produces a series of gametes in which the factors are coupled in this way. The intensity of the coupling, however, varies in different cases. Where we are dealing with another, viz. fertility (F) and the dark axil (D), the experimental numbers accord with the view that the gametic series is here 15 FD : 1 Fd : 1 fD : 15 fd. The coupling is in this instance more intense. In the case of the erect standard (E) and blueness (B) the coupling is even more intense, and the experimental evidence available at present points to the gametic series here being 63 *Eb* : 1 *EB* : 1 *eB* : 63 *eb*. There is evidence also for supposing that the intensity of the coupling may vary in different families for the same pair of factors. The coupling between blue and long pollen is generally on the 7 : 1 : 1 : 7 {97}basis, but in some cases it may be on the 15 : 1 : 1 : 15 basis. But though the intensity of the coupling may vary it varies in an orderly way. If A and B are the two factors concerned, the results obtained in F_2 are explicable on the assumption that the ratio of the four sorts of gametes produced is a term of the series —

3 *AB* +	*Ab* +	*aB* +	3 *ab*
7 *AB* +	*Ab* +	*aB* +	7 *ab*
15 *AB* +	*Ab* +	*aB* +	15 *ab*, etc., etc.

In such a series the number of gametes containing A is equal to the number lacking A, and the same is true for B. Consequently the number of zygotes formed containing A is three times as great as the number of zygotes which do not contain A; and similarly for B. The proportion of dominants to recessives in each case is 3 : 1. It is only in the distribution of the characters with relation to one another that these cases differ from a simple Mendelian case.

As the study of these series presents another feature of some interest, we may consider it in a little more detail. In the accompanying table are set out the results produced by these different series of gametes. The series marked by an asterisk have already been demonstrated experimentally. The first term in the series, {98}in which all the four kinds of gametes are produced in equal numbers is, of course, that of a simple Mendelian case where no coupling occurs.

No. of Gametes in series.	Distribution of Factors in Gametic Series AB. Ab. aB. ab.	No. of Zygotes produced.	Form of F_2 Generation.			
			AB.	Ab.	aB.	ab.
4	1 : 1 : 1 : 1	16	9	3	3	1
8	3 : 1 : 1 : 3	64	49	7	7	9

16	7 : 1 : 1 : 7	256	177	15	15	49*
32	15 : 1 : 1 : 15	1024	737	31	31	225*
64	31 : 1 : 1 : 31	4096	3009	63	63	961
128	63 : 1 : 1 : 63	16384	12161	127	127	3969*
$2n$	$(n-1) : 1 : 1 : (n-1)$	$4n^2$	$3n^2-(2n-1)$	$(2n-1)$	$(2n-1)$	$n^2-(2n-1)$

Now, as the table shows, it is possible to express the gametic series by a general formula $(n + 1) AB + Ab + aB + (n - 1) ab$, where $2n$ is the total number of the gametes in the series. A plant producing such a series of gametes gives rise to a family of zygotes in which $3n^2 - (2n - 1)$ show both of the dominant characters and $n^2 - (2n - 1)$ show both of the recessive characters, while the number of the two classes which each show one of the two dominants is $(2n - 1)$. When in such a series the coupling becomes closer the value of n increases, but in comparison with n^2 its value becomes less and less. The larger n becomes the more negligible is its value relatively to n^2. If, therefore, the coupling were very close, the series $3n^2 - (2n - 1) : (2n - 1) : (2n - 1) : n^2 - (2n - 1)$ would approximate more and more to the series $3n^2 : n^2$, i.e. to a simple 3 : 1 ratio. Though the point is probably of more theoretical than practical interest, it is not impossible that some of the cases which have hitherto been regarded as following a simple 3 : 1 ratio will turn out on further analysis to belong to this more complicated scheme.

{99}

CHAPTER X

SEX

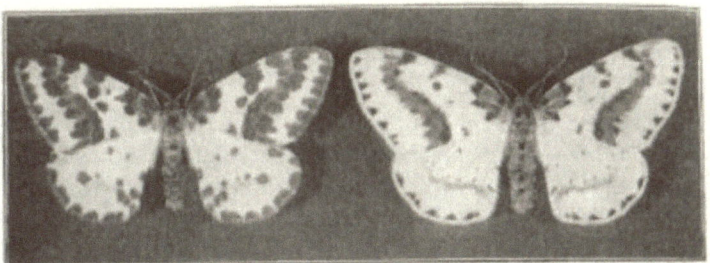

Fig. 17.

Abraxas grossulariata, the common currant moth, and (on the right) its paler lacticolor variety.

In their simplest expression the phenomena exhibited by Mendelian characters are sharp and clean cut. Clean cut and sharp also are the phenomena of sex. It was natural, therefore, that a comparison should have been early instituted between these two sets of phenomena. As a general rule, the cross between a male and a female results in the production of the two sexes in approximately equal numbers. The cross between a heterozygous dominant and a recessive also leads to equal numbers of recessives and of heterozygous dominants. Is it not, therefore, possible that one of the sexes is heterozygous for a factor which is lacking in the other, and that the presence or absence of this factor determines the sex of the zygote? The results of some recent experiments would appear to justify this interpretation, at any rate in particular cases. Of these, the simplest is that of the common currant moth (*Abraxas grossulariata*), of which there exists a pale variety (Fig. 17) known as *lacticolor*. The experiments of Doncaster and Raynor showed that the variety behaved as a simple recessive to the normal form. But the distribution of the dominants and {100}recessives

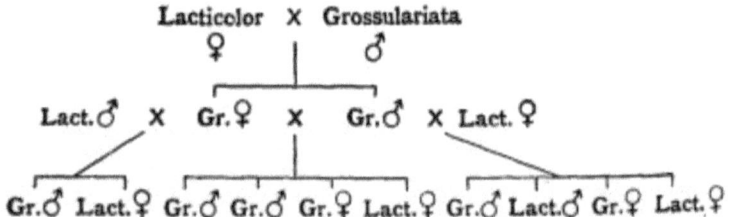

with regard to the sexes was peculiar. The original cross was between a *lacticolor* female and a normal male. All the F₁ moths of both sexes were of the normal *grossulariata* type. The F₁ insects were then paired together and gave a generation consisting of 3 normals : 1 *lacticolor*. But all the *lacticolor* were females, and all the males were of the normal pattern. It was, however, found possible to obtain the *lacticolor male* by mating a *lacticolor* female with the F₁ male. The family resulting from this cross consisted of normal males and normal females, *lacticolor* males and *lacticolor* females, and the {101}four sorts were produced in approximately equal numbers. In such a family there was no special association of either of the two colour varieties with one sex rather than the other. But the reverse cross, F₁ female by *lacticolor* male, gave a very different result. As in the previous cross such families contained equal numbers of the normal form and of the recessive variety. But all of the normal *grossulariata* were males, while all the *lacticolor* were females. Now this seemingly complex collection of facts is readily explained if we make the following three assumptions: —

(1) The *grossulariata* character (G) is dominant to the lacticolor character (g). This is obviously justified by the experiments, for, leaving the sex distribution out of account, we get the expected 3 : 1 ratio from F₁ × F₁, and also the expected ratio of equality when the heterozygote is crossed with the recessive.

(2) The female is heterozygous for a dominant factor (F) which is lacking in the male. The constitution of a female is consequently Ff, and of a male ff. This assumption is in harmony with the fact that the sexes are produced in approximately equal numbers.

(3) There exists repulsion between the factors G and F in a zygote which is heterozygous for them both. Such zygotes (FfGg) must

always be females, and on this assumption will produce gametes Fg and fG in equal numbers. {102}

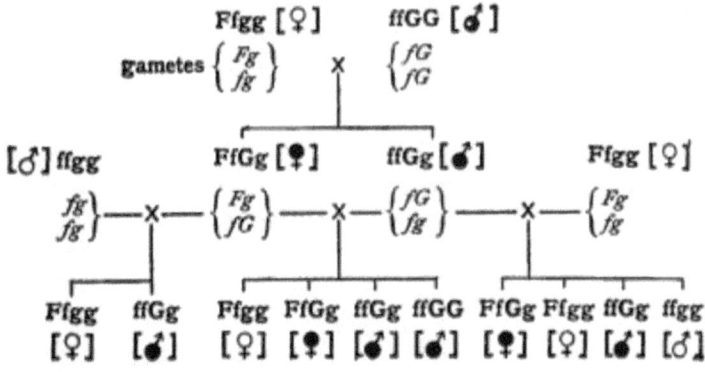

Fig. 18.

> Scheme of inheritance in the F_1 and F_2 generations resulting from the cross of *lacticolor* female with *grossulariata* male. The character of each individual is represented by the sex signs in brackets, the black being *grossulariata* in appearance and the light ones *lacticolor*.

We may now construct a scheme for comparison with that on page 100 to show how these assumptions explain the experimental results. The original parents were *lacticolor* female and *grossulariata* male, which on our assumptions must be $Ffgg$ and $ffGG$ respectively in constitution. Since the female is always heterozygous for F, her gametes must be of two kinds, viz. Fg and fg, while those of the pure *grossulariata* male must be all fG. When an ovum Fg is fertilised by a spermatozoon fG, the resulting zygote, $FfGg$, is heterozygous for both F and G, and in appearance is a female *grossulariata*. The zygote resulting from the fertilisation of an ovum fg by a spermatozoon fG is heterozygous for G, but does not contain F, and therefore is a male *grossulariata*. Such a male being in constitution {103} $ffGg$ must produce gametes of two kinds, fG and fg, in equal numbers. And since we are assuming repulsion between F and G, the F_1 female being in constitution $FfGg$, must produce equal numbers of gametes Fg and fG. For on our assumption F and G cannot enter into the same gamete. The series of gametes produced by the F_1 moths, therefore, are *fG*, *fg* by the male and *Fg*, *fG* by the female. The resulting F_2 generation consequently consists of the four classes of zygotes *Ffgg*, *FfGg*, *ffGg*, and *ffGG* in equal numbers. In other words, the sexes are produced in equal numbers, the proportion of normal grossulariata to *lacticolor* is 3 : 1, and all of the *lacticolor* are females; that is to say, the results worked out on our assumptions accord with those actually produced by experiment. We may now turn to the results which should be obtained by crossing the F_1 moths with the *lacticolor* variety. And first we will take the cross *lacticolor* female × F_1 male. The gametes produced by the lacticolor female we have already seen to be *Fg* and *fg*, while those produced by the F_1 male are *fG* and *fg*. The bringing together of these two series of gametes must result in equal numbers of the four kinds of zygotes *FfGg*, *Ffgg*, *ffGg*, and *ffgg*, *i.e.* of female *grossulariata* and *lacticolor*, and of male *grossulariata* and *lacticolor* in equal numbers. Here, again, the calculated results accord with those of experiment. Lastly, we may examine what should happen when the F_1 female is crossed with the *lacticolor* {104}male. The F_1 female, owing to the repulsion between *F* and *G*, produces only the two kinds of ova *Fg* and *fG*, and produces them in equal numbers. Since the *lacticolor* male can contain neither *F* nor *G*, all of its spermatozoa must be *fg*. The results of such a cross, therefore, should be to produce equal numbers of the two kinds of zygote *Ffgg* and *ffGg*, *i.e.* of *lacticolor* females and of *grossulariata* males. And this, as we have already seen, is the actual result of such a cross.

Before leaving the currant moth we may allude to an interesting discovery which arose out of these experiments. The *lacticolor* variety in Great Britain is a southern form and is not known to occur in Scotland. Matings were made between wild Scotch females and *lacticolor* males. The families resulting from such matings were precisely the same as those from *lacticolor* males and F_1 females, viz. *grossulariata* males and *lacticolor* females only. We are, therefore, forced to regard the constitution of the wild *grossulariata* female as identical with that of the F_1 female, *i.e.* as heterozygous for the *grossulariata* factor as well as for the factor for femaleness. Though from a region where *lacticolor* is unknown, the "pure" wild *grossulariata* female is nevertheless a permanent mongrel, but it can never reveal its true colours unless it is mated with a male which is either heterozygous for *G* or pure *lacticolor*. And as all the wild northern males are {105}pure for the *grossulariata* character this can never happen in a state of nature.

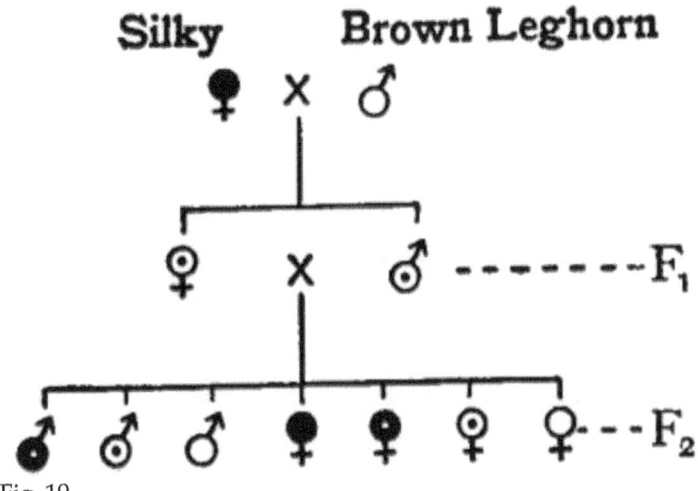

Fig. 19.

Scheme illustrating the result of crossing a Silky hen with a Brown Leghorn cock. Black sex signs denote deeply pigmented birds, and light sex signs those without pigmentation. The light signs with a black dot in the centre denote birds with a small amount of pigment.

An essential feature of the case of the currant moth lies in the different results given by reciprocal crosses. *Lacticolor* female × *grossulariata* male gives *grossulariata* alone of both sexes. But *grossulariata* female × *lacticolor* male gives only *grossulariata* males and *lacticolor* females. Such a difference between reciprocal crosses has also been found in other animals, and the experimental results, though sometimes more complicated, are explicable on the same lines. An interesting case in which three factors are concerned has been recently worked out in poultry. The Silky breed of fowls is characterised among other peculiarities by a remarkable abundance of melanic pigment. The skin is dull black, while the comb and wattles are of a deep purple colour contrasting sharply with the white plumage (Pl. V., 3). Dissection shows that this black pigment is widely spread throughout the body, being especially marked in such membranes as the mesenteries, the periosteum, and the pia mater surrounding the brain. It also occurs in the connective tissues among the muscles. In the Brown Leghorn, on the other hand, this pigment is not found. Reciprocal crosses between these two breeds gave a remarkable difference in result. A cross between the Silky hen and the Brown Leghorn cock produced F_1 birds in which both sexes exhibited only traces of the pigment. On casual observation they might have {106}passed for unpigmented birds, for with the exception of an occasional fleck of pigment their skin, comb and wattles were as clear as in the Brown Leghorn (Pl. V., 1 and 4). Dissection revealed the presence of a slight amount of internal pigment. Such birds bred together gave some offspring with the full pigmentation of the Silky, some without any pigment, and others showing different degrees of pigment. None of the F_2 male birds, however, showed the full deep pigmentation of the Silky.

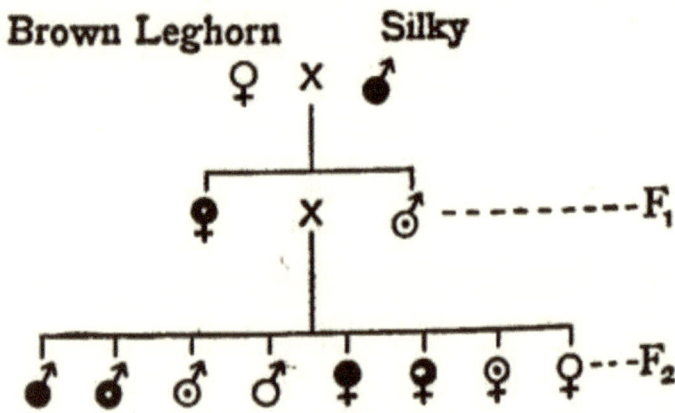

Fig. 20.

Scheme illustrating the result of crossing a Brown Leghorn hen with a Silky cock (cf. Fig. 19).

When, however, the cross was made the other way, viz. Brown Leghorn hen × Silky cock, the result was different. While the F_1

male birds were almost destitute of pigment as in the previous cross, the F_1 hens, on the other hand, were nearly as deeply pigmented as the pure Silky {107}(Pl. V., 2). The male Silky transmitted the pigmentation, but only to his daughters. Such birds bred together gave an F_2 generation containing chicks with the full deep pigment, chicks without pigment, and chicks with various grades of pigmentation, all the different kinds in both sexes.

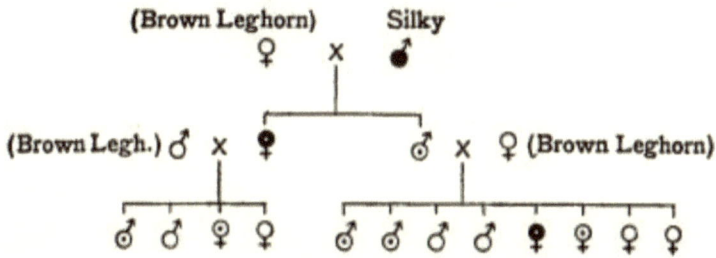

Fig. 21.

Scheme to illustrate the result of crossing F_1 birds (*e.g.* Brown Leghorn × Silky) with the pure Brown Leghorn.

In analysing this complicated case many other different crosses were made, but for the present it will be sufficient to mention but one of these, viz. that between the F_1 birds and the pure Brown Leghorn. The cross between the F_1 hen and the Brown Leghorn cock produced only birds with a slight amount of pigment and birds without pigment. And this was true for both the deeply pigmented and the slightly pigmented types of F_1 hen. But when the F_1 cock was mated to a Brown Leghorn hen, a definite proportion of the chicks, one in eight, was deeply pigmented, and *these deeply pigmented birds were always females* (cf. Fig. 21). And in this respect all the F_1 males behaved alike, whether they were from the Silky hen or from the Silky cock. We have, therefore, the paradox that the F_1 hen, though herself deeply pigmented, cannot transmit this condition to any of her offspring when she is mated to the unpigmented Brown Leghorn, but that, when similarly mated, the F_1 cock can transmit this pigmented condition to a quarter of his female offspring though he himself is almost devoid of pigment.

Plate V.

1, 2, F_1 Cock and Hen, ex Brown Leghorn Hen × Silky Cock; 3, Silky Cock; 4, Hen ex Silky Hen × Brown Leghorn Cock.

{108}

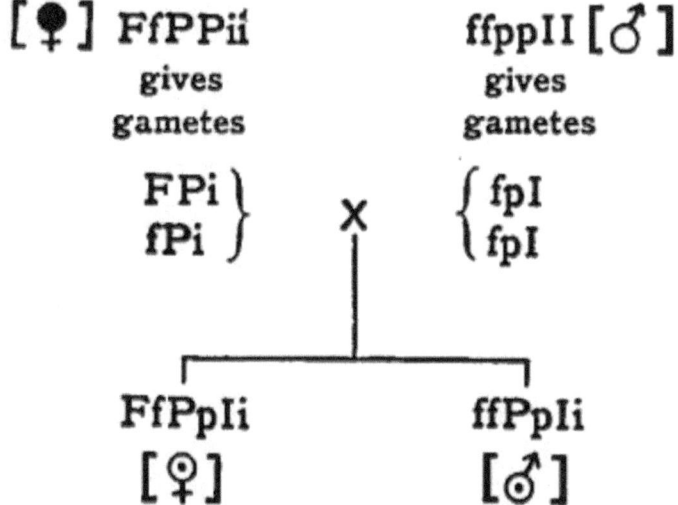

Fig. 22.

> Scheme to illustrate the nature of the F_1 generation from the Silky hen and Brown Leghorn cock (cf. Fig. 23).

Now all these apparently complicated results, as well as many others to which we have not alluded, can be expressed by the following simple scheme. There are three factors affecting pigment, viz. (1) a pigmentation factor (P); (2) a factor which inhibits the production of pigment (I); and (3) a factor for femaleness (F), for which the female birds are heterozygous, but which is not present in the males. Further, we make the assumptions (*a*) that there is repulsion between F and I in the female zygote ($FfIi$), and (*b*) that the male Brown Leghorn is homozygous for the inhibitor factor (I), but that the hen Brown Leghorn is always heterozygous for this factor just in the same way as the female of the currant moth is always heterozygous for the *grossulariata* factor. We may now proceed to show how this explanation fits the experimental facts which we have given.

The Silky is pure for the pigmentation factor, but does not contain the inhibitor factor. The Brown Leghorn, on the other hand, contains the inhibitor factor, but not the {109}pigmentation factor. In crossing a Silky hen with a Brown Leghorn cock we are mating two birds of the constitution *FfPPii* and *ffppII*, and all the F₁ birds are consequently heterozygous for both *P* and *I*. In such birds the pigment is almost but not completely suppressed, and as both sexes are of the same constitution with regard to these two factors they are both of similar appearance.

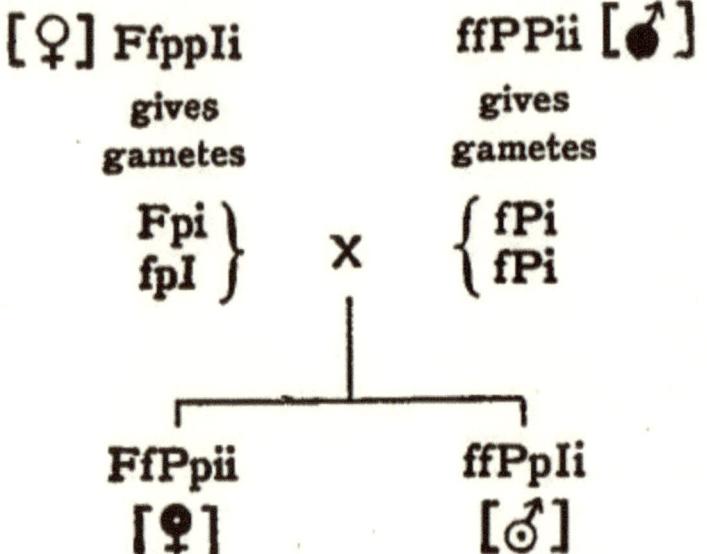

Fig. 23.

> Scheme to illustrate the nature of the F₁ generation from the Brown Leghorn hen and Silky cock (cf. Fig. 22).

In the reciprocal cross, on the other hand, we are mating a Silky male (*ffPPii*) with a Brown Leghorn hen which on our assumption is heterozygous for the inhibitor factor (*I*), and in constitution therefore is *FfppIi*. Owing to the repulsion between *F* and *I* the gametes produced by such a bird are *Fpi* and *fpI* in equal numbers. All the gametes produced by the Silky cock are *fPi*. Hence the constitution of the F_1 male birds produced by this cross is *ffPpIi* as before, but the female birds must be all of the constitution *FfPpii*. The Silky cock transmits the fully pigmented condition to his daughters, because the gametes of the Brown Leghorn hen which contain the factor for femaleness do not contain the {110}inhibitory factor owing to the repulsion between these factors. The nature of the F_2 generation in each case is in harmony with the above scheme. As, however, it serves to illustrate certain points in connection with intermediate forms we shall postpone further consideration of it till we discuss these matters, and for the present shall limit ourselves to the explanation of the different behaviour of the F_1 males and females when crossed with the Brown Leghorn. And, first, the cross of Brown Leghorn female by F_1 male. The Brown Leghorn hen is on our hypothesis *FfppIi*, and produces gametes *Fpi* and *fpI*. The F_1 cock is on our hypothesis *ffPpIi*, and produces in equal numbers the four kinds of gametes *fPI*, *fPi*, *fpI*, *fpi*. The result of the meeting of these two series of gametes is given in Fig. 24. Of the eight different kinds of zygote formed only one contains *P* in the absence of *I*, and this is a female. The result, as we have already seen, is in accordance with the experimental facts.

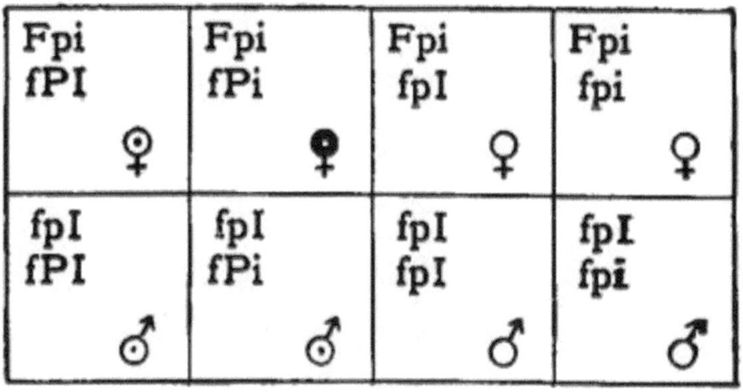

Fpi fPI ♀	Fpi fPi ♀	Fpi fpI ♀	Fpi fpi ♀
fpI fPI ♂	fpI fPi ♂	fpI fpI ♂	fpI fpi ♂

Fig. 24.

Diagram showing the nature of the offspring from a Brown Leghorn hen and an F_1 cock bred from Silky hen × Brown Leghorn cock, or *vice versa*.

On the other hand, the Brown Leghorn cock is on our hypothesis *ffppII*. All his gametes consequently contain the inhibitor factor, and when he is mated with an F_1 {111}hen all the zygotes produced must contain *I*. None of his offspring, therefore, can be fully pigmented, for this condition only occurs in the absence of the inhibitor factor among zygotes which are either homozygous or heterozygous for *P*.

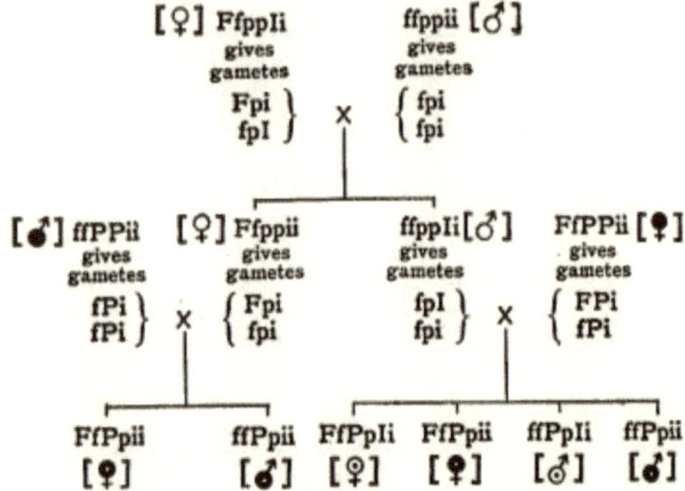

Fig. 25.

Scheme to illustrate the heterozygous nature of the pure Brown Leghorn hen. For explanation see text.

The interpretation of this case turns upon the constitution of the Brown Leghorn hen, upon her heterozygous condition with regard to the two factors *F* and *I*, and upon the repulsion that occurs between them when the gametes are formed. Through an independent set of experiments this view of the nature of the Brown Leghorn hen has been confirmed in an interesting way. There are fowls which possess neither the factor for pigment nor the inhibitory factor, which are in constitution *ppii*. Such birds when crossed with the Silky give dark pigmented birds of both sexes in F_1, and the F_2 generation consists of pigmented and unpigmented in the ratio 3 : 1. Now a cock of such a strain crossed with a Brown Leghorn hen should give only completely unpigmented birds. But if, as we have supposed, the Brown Leghorn hen is producing gametes *Fpi* and *fpI*, the male birds produced by such a cross should be heterozygous for *I*, {112} *i.e.* in constitution *ffppIi*, while the hen birds, though identical in appearance so far as absence of pigmentation goes, should not contain this factor but should be constitutionally *Ffppii*. Crossed with the pure Silky, the F_1 birds of opposite sexes should give an entirely different result. For while the hens should give only deeply pigmented birds of both sexes, the cocks should give equal numbers of deeply pigmented and slightly pigmented birds (cf. Fig. 25). These were the results which the experiment actually gave, thus affording strong confirmation of the view which we have been led to take of the Brown Leghorn hen. Essentially the poultry case is that of the currant moth. It differs in that the factor which {113}repels femaleness produces no visible effect, and its presence or absence can only be determined by the introduction of a third factor, that for pigmentation.

This conception of the nature of the Brown Leghorn hen leads to a curious paradox. We have stated that the Silky cock transmits the pigmented condition, but transmits it to his daughters only. Apparently the case is one of unequal transmission by the father. Actually, as our analysis has shown, it is one of unequal transmission by the mother, the father's contribution to the offspring being identical for each sex. The mother transmits to the daughters her dominant quality of femaleness, but to balance this, as it were, she transmits to her sons another quality which her daughters do not receive. It is a matter of common experience among human families that in respect to particular qualities the sons tend to resemble their mothers more than the daughters do, and it is not improbable that such observations have a real foundation for which the clue may be provided by the Brown Leghorn hen.

Nor is this the only reflection that the Brown Leghorn suggests. Owing to the repulsion between the factors for femaleness and for pigment inhibition, it is impossible by any form of mating to make a hen which is homozygous for the inhibitor factor. She has bartered away for femaleness the possibility of ever receiving a double dose of this factor. We know that in some cases, as, for example, {114}that of the blue Andalusian fowl, the qualities of the individual are markedly different according as to whether he or she has received a single or a double dose of a given factor. It is not inconceivable that some of the qualities in which a man differs from a woman are founded upon a distinction of this nature. Certain qualities of intellect, for example, may depend upon the existence in the individual of a double dose of some factor which is repelled by femaleness. If this is so, and if woman is bent upon achieving the results which such qualities of intellect imply, it is not education or training that will help her. Her problem is to get the factor on which the quality depends into an ovum that carries also the factor for femaleness.

{115}

CHAPTER XI

SEX (*continued*)

The cases which we have considered in the last chapter belong to a group in which the peculiarities of inheritance are most easily explained by supposing that the female is heterozygous for some factor that is not found in the male. Femaleness is an additional character superposed upon a basis of maleness, and as we imagine that there is a separate factor for each the full constitutional formula for a female is *FfMM*, and for a male *ffMM*. Both sexes are homozygous for the male element, and the difference between them is due to the presence or absence of the female element *F*.

There are, however, other cases for which the explanation will not suffice, but can be best interpreted on the view that the male is heterozygous for a factor which is not found in the female. Such a case is that recently described by Morgan in America for the pomace fly (*Drosophila ampelophila*). Normally this little insect has a red eye, but white eyed individuals are known to occur as rare sports. Red eye is dominant to white. In their relation to sex the eye colours of the pomace fly {116}are inherited on the same lines as the *grossulariata* and *lacticolor* patterns of the currant moth, but with one essential difference. The factor which repels the red-eye factor is in this case to be found in the male, and here consequently it is the male which must be regarded as heterozygous for a sex factor that is lacking in the female.

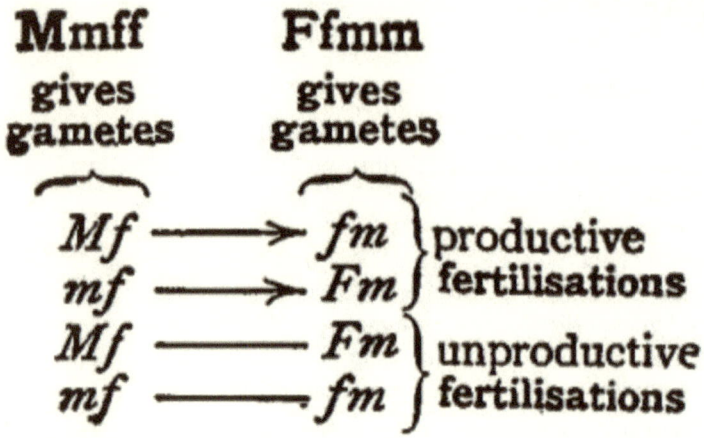

In order to bring these cases and others into line an interesting suggestion has recently been put forward by Bateson. On this suggestion each sex is heterozygous for its own sex factor only, and does not contain the factor proper to the opposite sex. The male is of the constitution, *Mmff* and the female *Ffmm*. Each sex produces two sorts of gametes, *Mf* and *mf* in the case of the male, and *Fm, fm* in that of the female. But on this view a further supposition is necessary. If each of the two kinds of spermatozoa were capable of fertilising each of the two kinds of ova, we should get individuals of the constitution *MmFf* and *mmff*, as well as the normal males and females, *Mmff* and *Ffmm*. As the facts of ordinary bisexual reproduction afford us no grounds for assuming the existence of these two classes of individuals, whatever they may be, we must suppose that fertilisation. is productive only between the spermatozoa carrying *M* and the ova without *F*, or between the spermatozoa {117}without *M* and the ova containing *F*. In other words we must on this view suppose that fertilisations between certain forms of gametes, even if they can occur, are incapable of giving rise to zygotes with the capacity for further development. If we admit this supposition, the scheme just given will cover such cases as those of the currant moth and the fowl, equally as well as that of the pomace fly. In the former there is repulsion between either the *grossulariata* factor and *F*, or else between the pigment inhibitor factor and *F*, while in the latter there is repulsion between the factor for red eye and *M*.

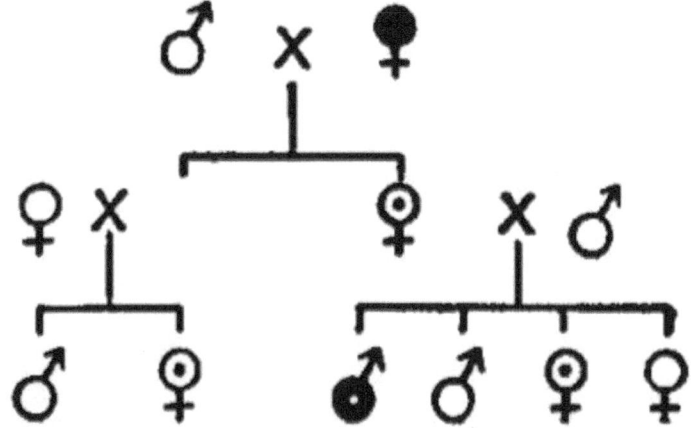

Fig. 26.

> Scheme to illustrate the probable mode of inheritance of colour-blindness. The dark signs represent affected individuals. A black dot in the centre denotes an unaffected female who is capable of transmitting the condition to her sons.

Whatever the merits or demerits of such a scheme it certainly does offer an explanation of a peculiar form of sex limited inheritance in man. It has long been a matter of common knowledge that colour-blindness is much more common among men than among women, and also that unaffected women can transmit it to their sons. At first sight the case is not unlike that of the sheep, where the horned character is apparently dominant in the male but recessive in the female. The hypothesis that the colour-blind condition is due to the presence of an extra factor as compared with the normal, and that a single dose of it will produce {118}colour-blindness in the male but not in the female, will cover a good many of the observed facts (cf. Fig. 26). Moreover, it serves to explain the remarkable fact that *all* the sons of colour-blind women are also colour-blind. For a woman cannot be colour-blind unless she is homozygous for the colour-blind factor, in which case all her children must get a single dose of it even if she marries a normal male. And this is sufficient to produce colour-blindness in the male, though not in the female.

But there is one notable difference in this case as compared with that of the sheep. When crossed with pure hornless ewes the heterozygous horned ram transmits the horned character to half his male offspring (cf. p. 71). But the heterozygous colour-blind man does not behave altogether like a sheep, for he apparently does not transmit the colour-blind condition to any of his male offspring. If, however, we suppose that the colour-blind factor is repelled by the factor for maleness, the amended scheme will cover the observed facts. For, denoting the colour-blind factor by X, the gametes produced by the colour-blind male are of two sorts only, viz. *Mfx* and *mfX*. If he marries a normal woman (*Ffmmxx*), the spermatozoa *Mfx* unite with ova *fmx* to give normal males, while the spermatozoa *mfX* unite with ova *Fmx* to give females which are heterozygous for the colour-blind factor. These daughters are themselves normal, but transmit the condition to about half their sons. {119}

The attempt to discover a simple explanation of the nature of sex has led us to assume that certain combinations between gametes are incapable of giving rise to zygotes which can develop further. In the various cases hitherto considered there is no reason to suppose that anything of the sort occurs, or that the different gametes are otherwise than completely fertile one with another. One peculiar case, however, has been known for several years in which some of the gametes are apparently incapable of uniting to produce offspring. Yellow in the mouse is dominant to agouti, but hitherto a homozygous yellow has never been met with. The yellows from families where only yellows and agoutis occur produce, when bred together, yellows and agoutis in the ratio 2 : 1. If it were an ordinary Mendelian case the ratio should be 3 : 1, and one out of every three yellows so bred should be homozygous and give only yellows when crossed with agouti. But Cuénot and others have shown that *all* of the yellows are heterozygous, and when crossed with agoutis give both yellows and agoutis. We are led, therefore, to suppose that an ovum carrying the yellow factor is unproductive if fertilised by a spermatozoon which also bears this factor. In this way alone does it seem possible to explain the deficiency of yellows and the absence of homozygous ones in the families arising from the mating of yellows together. At present, however, it remains the only definite instance among animals in which we have {120}grounds for assuming that anything in the nature of unproductive fertilisation takes place. [8]

If we turn from animals to plants we find a more complicated state of affairs. Generally speaking, the higher plants are hermaphrodite, both ovules and pollen grains occurring on the same flower. Some plants, however like most animals, are of separate sexes, a single plant bearing only male or female flowers. In other plants the separate flowers are either male or female, though both are borne on the same individual. In others, again, the conditions are even more complex, for the same plant may bear flowers of three kinds, viz. male, female, and hermaphrodite. Or it may be that these three forms occur in the same species but in different individuals — female and hermaphrodites in one species; males, females, and hermaphrodites in another. One case, however, must be mentioned as it suggests a possibility which we have not hitherto encountered. In the common English bryony (*Bryonia dioica*) the sexes are separate, some plants having only male and others only female flowers. In another European species, *B. alba*, both male and female flowers occur on the same plant. Correns crossed these two species reciprocally, and also fertilised *B. dioica* by its own male with the following results: —

{121}

dioica ♀	× dioica	♂ gave	♀♀ and ♂♂
"	× alba	♂ "	♀♀ only
alba ♀	× dioica	♂ "	♀♀ and ♂♂.

The point of chief interest lies in the striking difference shown by the reciprocal crosses between *dioica* and *alba*. Males appear when *alba* is used as the female parent but not when the female *dioica* is crossed by male *alba*. It is possible to suggest more than one scheme to cover these facts, but we may confine ourselves here to that which seems most in accord with the general trend of other cases. We will suppose that in *dioica* femaleness is dominant to maleness, and that the female is heterozygous for this additional factor. In this species, then, the female produces equal numbers of ovules with and without the female factor, while this factor is absent in all the pollen grains. *Alba* ♀ × *dioica* ♂ gives the same result as *dioica* ♀ × *dioica* ♂, and we must therefore suppose that alba produces male and female ovules in equal numbers. *Alba* ♂ × *dioica* ♀, however, gives nothing but females. Unless, therefore, we assume that there is selective fertilisation we must suppose that all the pollen grains of alba carry the female factor—in other words, that so far as the sex factors are concerned there is a difference between the ovules and pollen grains borne by the same plant. Unfortunately further investigation of this case is rendered impossible owing to the complete sterility of the F_1 plants. {122}

Fig. 27.

Single and double stocks raised from the same single parent.

That the possibility of a difference between the ovules and pollen grains of the same individual must be taken into account in future work there is evidence from quite a different source. The double stock is an old horticultural favourite, and for centuries it has been known that of itself it sets no seed, but must be raised from special strains of the single variety. "You must understand withall," wrote John Parkinson of his gilloflowers, [9] "that those plants that beare double flowers, doe beare

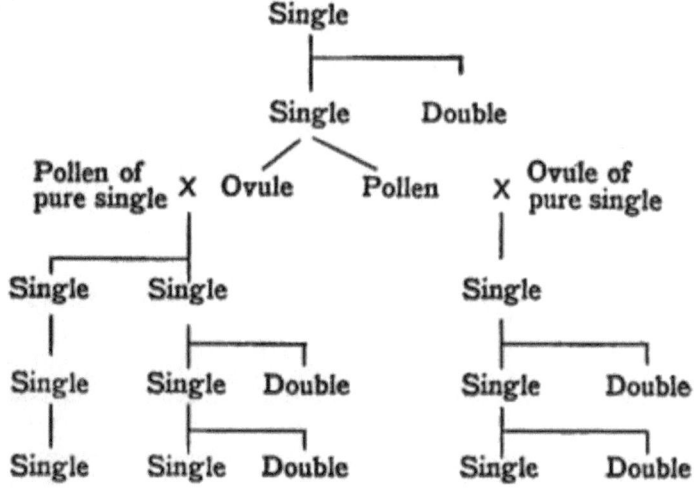

no seed at all ... but the onely way to have double flowers any yeare is to save the seedes of those plants of this kinde that beare single flowers, for from that seede will rise some that will beare single, and some double flowers." With regard to the nature of these double-throwing strains of singles, Miss Saunders has recently brought out some interesting facts. She crossed the double-throwing singles with pure singles belonging to strains in which doubles never occur. The cross was made both ways, and in both cases all the F_1 plants were single. A distinction, however, appeared when a further generation was raised from the F_1 plants. All the F_1 plants from the pollen of the double-throwing single behaved like double-throwing singles, but of the F_1 plants from the ovules of the double throwers some behaved as double throwers, and some as pure singles. We are led to infer, therefore, that the ovules and pollen grains {123}of the double throwers, though both produced by the same plant, differ in their relation to the factor (or factors) for doubleness. Doubleness is apparently carried by all the pollen grains of such plants, but only by some of the ovules. Though the nature of doubleness in stocks is not yet clearly understood, the facts discovered by Miss Saunders suggest strongly that the ovules and pollen grains of the same plant may differ in their transmitting properties, probably owing to some process of segregation in the growing plant which leads to an unequal distribution of some or other factors to the cells which give rise to the ovules as compared with those from which {124}the pollen grains eventually spring. Whether this may turn out to be the true account or not, the possibility must not be overlooked in future work.

From all this it is clear enough that there is much to be done before the problem of sex is solved even so far as the biologist can ever expect to solve it. The possibilities are many, and many a fresh set of facts is needed before we can hope to decide among them. Yet the occasional glimpses of clear-cut and orderly phenomena, which Mendelian spectacles have already enabled us to catch, offer a fair hope that some day they may all be brought into focus, and assigned their proper places in a general scheme which shall embrace them all. Then, though not till then, will the problem of the nature of sex pass from the hands of the biologist into those of the physicist and the chemist.

{125}

CHAPTER XII

INTERMEDIATES

So far as we have gone we have found it possible to express the various characters of animals and plants in terms of definite factors which are carried by the gametes, and are distributed according to a definite scheme. Whatever may be the nature of these factors it is possible for purposes of analysis to treat them as indivisible entities which may or may not be present in any given gamete. When the factor is present it is present as a whole. The visible properties developed by a zygote in the course of its growth depend upon the nature and variety of the factors carried in by the two gametes which went to its making, and to a less degree upon whether each factor was brought in by both gametes or by one only. If the given factor is brought in by one gamete only, the resulting heterozygote may be more or less intermediate between the homozygous form with a double dose of the factor and the homozygous form which is entirely destitute of the factor. Cases in point are those of the primula flowers and the Andalusian fowls. Nevertheless these intermediates produce only pure gametes, as is {126}shown by the fact that the pure parental types appear in a certain proportion of their offspring. In such cases as these there is but a single type of intermediate, and the simple ratio in which this and the two homozygous forms appear renders the interpretation obvious. But the nature of the F_2 generation may be much more complex, and, where we are dealing with factors which interact upon one another, may even present the appearance of a series of intermediate forms grading from the condition found in one of the original parents to that which occurred in the other. As an illustration we may consider the cross between the Brown Leghorn and Silky fowls which we have already dealt with in connection with the inheritance of sex. The offspring of a Silky hen mated with a Brown Leghorn are in both sexes birds with but a trace of the Silky pigmentation. But when such birds are bred together they produce a generation consisting of chicks as deeply pigmented as the original Silky parent, chicks devoid of pigment like the Brown Leghorn, and chicks in which the pigmentation shows itself in a variety of intermediate stages. Indeed from a hundred chicks bred in this way it would be possible to pick out a number of individuals and arrange them in an apparently continuous series of gradually increasing pigmentation, with the completely unpigmented at one end and the most deeply pigmented at the other. Nevertheless, the case is one in which complete segregation of the different factors takes {127}

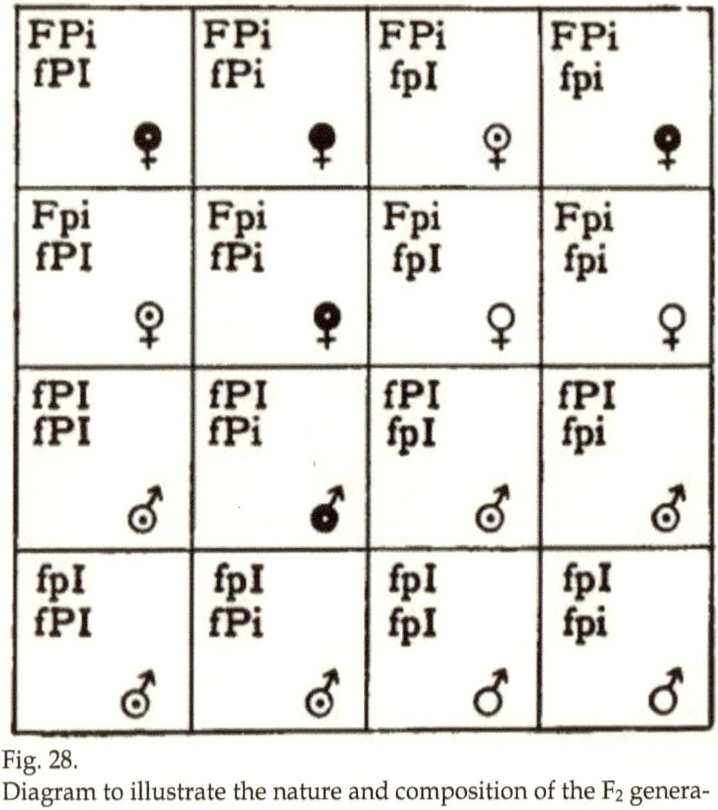

FPi fPI ♀	FPi fPi ♀	FPi fpI ♀	FPi fpi ♀
Fpi fPI ♀	Fpi fPi ♀	Fpi fpI ♀	Fpi fpi ♀
fPI fPI ♂	fPI fPi ♂	fPI fpI ♂	fPI fpi ♂
fpI fPI ♂	fpI fPi ♂	fpI fpI ♂	fpI fpi ♂

Fig. 28.
Diagram to illustrate the nature and composition of the F₂ generations arising from the cross of Silky hen with Brown Leghorn cock.place, and the apparently continuous series of intermediates is the result of the interaction of the different factors upon one another. The constitution of the F₁ ♂ is a *ffPpIi*, and such a bird produces in equal numbers the four sorts of gametes *fPI, fPi, fpI, fpi*. The constitution of the F₁ ♀ in this case is *FfPpIi*. Owing to the repulsion between F and I she produces the four kinds of gametes *FPi, Fpi, fPI, fpi*, and produces them in equal numbers. The result of bringing two

such series of gametes together is shown in Fig. 28. Out of the sixteen types of zygote formed one (*FfPPii*) is homozygous for the pigmentation factor, and does not contain the inhibitor factor. Such a bird is as deeply pigmented as the pure Silky parent. Two, again, contain a single dose of *P* in the absence of *I*. These are nearly as dark as the pure Silky. Four zygotes are destitute of *P*, though they may or may not contain *I*. These birds are completely devoid of pigment like the Brown Leghorn. The remaining nine zygotes show {128}various combinations of the two factors *P* and *I*, being either *PPIi*, *PPII*, *PpII*, or *PpIi*, and in each of these cases the pigment is more or less intense according to the constitution of the bird. Thus a bird of the constitution *PPIi* approaches in pigmentation a bird of the constitution *Ppii*, while a bird of the constitution *PpII* has but little more pigment than the unpigmented bird. In this way we have seven distinct grades of pigmentation, and the series is further complicated by the fact that these various grades exhibit a rather different amount of pigmentation according as they occur in a male or a female bird, for, generally speaking, the female of a given grade exhibits rather more pigment than the corresponding male. The examination of a number of birds bred in this way might quite well suggest that in this case we were dealing with a character which could break up, as it were, to give a continuous series of intergrading forms between the two extremes. With the constant handling of large numbers it becomes possible to recognise most of the different grades, though even so it is possible to make mistakes. Nevertheless, as breeding tests have amply shown, we are dealing with but two interacting factors which segregate cleanly from one another according to the strict Mendelian rule. The approach to continuity in variation exhibited by the F_2 generation depends upon the fact that these two factors interact upon one another, and to different degrees according as the zygote is for one {129}or other or both of them in a homozygous or a heterozygous state. Moreover, certain of these intermediates will breed true to an intermediate condition of the pigmentation. A male of the constitution *ffPPII* when bred with females of the constitution *FfPPIi* will produce only males like itself and females like the maternal parent. We have dealt with this case in some detail, because the existence of families showing a series of intermediate stages between two characters has sometimes been brought forward in opposition to the view that the characters of organisms depend upon specific factors which are transmitted according to the Mendelian rule. But, as this case from poultry shows clearly, neither the existence of such a continuous series of intermediates, nor the fact that some of them may breed true to the intermediate condition, are incompatible with the Mendelian principle of segregation.

In connection with intermediates a more cogent objection to the Mendelian view is the case of the first cross between two definite varieties thenceforward breeding true. The case that will naturally occur to the reader is that of the mulatto, which results from the cross between the negro and the white. According to general opinion, these mulattos, of intermediate pigmentation, continue to produce mulattos. Unfortunately this interesting case has never been critically investigated, and the statement that the mulatto breeds true rests almost entirely upon {130}information that is general and often vague. It may be that the inheritance of skin pigmentation in this instance is a genuine exception to the normal rule, but at the same time it must not be forgotten that it may be one in which several interacting factors are concerned, and that the pure white and the pure black are the result of combinations which from their rarity are apt to be overlooked. But until we are in possession of accurate information it is impossible to pronounce definitely upon the nature of the inheritance in this case.

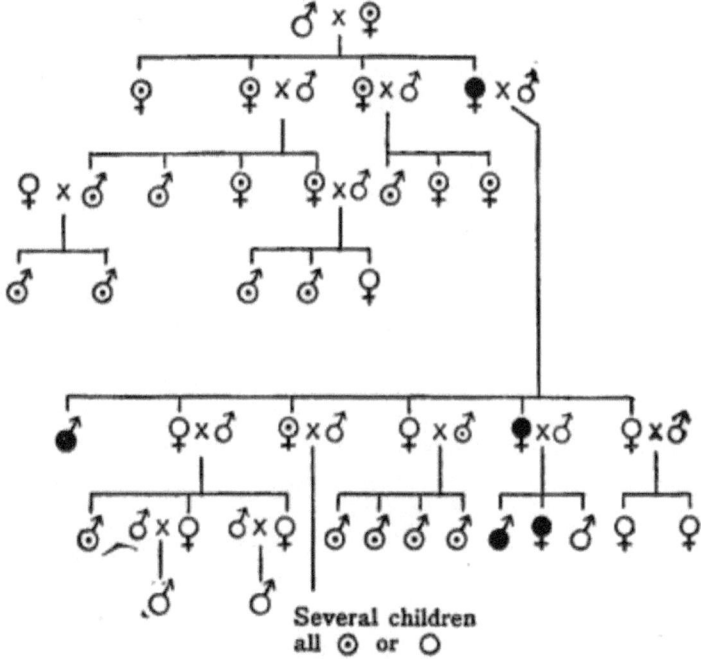

Fig. 29.

Pedigree of a family which originated from a cross between a Hindu and a European. Black signs denote individuals as dark as average Hindus. Plain signs denote quite-fair members, while those with a dot in the centre are intermediate.

{131}

On the other hand, from the cross between the darkly pigmented Eastern races and the white segregation seems to occur in subsequent generations. Families are to be found in which one parent is a pure white, while the other has arisen from the cross between the dark and light in the first or some subsequent generation. Such families may contain children indistinguishable from pure blonds as well as children of very dark and of intermediate shades. As an example, I may give the following pedigree, which was kindly communicated to me by an Anglo-Indian friend (Fig. 29). The family had resided in England for several generations, so that in this case there was no question of a further admixture of black. Most noticeable is the family produced by a very dark lady who had married a white man. Some of the children were intermediate in colour, but two were fair whites and two were dark as dark Hindus. This sharp segregation or splitting out of blacks and whites in addition to intermediates strongly suggests that the nature of the inheritance is Mendelian, though it may be complicated by the existence of several factors which may also react upon one another. Nor must it be forgotten that in so far as these different factors are concerned the whites themselves may differ in constitution without showing any trace of it in their appearance. Before the case can be regarded as settled all these different possibilities will have to be definitely tested. With the dark Eastern races as with the negro we cannot {132}hope to come to any conclusion until we have evidence collected by critical and competent observers.

Though for the present we must regard the case of the negro as not proven, there are nevertheless two others in which the heredity would appear not to follow the Mendelian rule. Castle in America crossed the lop-eared rabbit with the normal form, and found that the F_1 animals were intermediate with respect to their ears. And subsequent experiment showed that, on the whole, they bred true to this intermediate condition. The other case relates to Lepidoptera. The speckled wood butterfly (*Pararge egeria*) has a southern form which differs from the northern one in the greater brightness and depth of its yellow-brown markings. The northern form is generally distinguished as var. *egeriades*. Bateson crossed the southern form from the south of France with the paler British form, and found that the offspring were more or less intermediate in colour, and that in subsequent generations the parental types did not recur. These cases at present stand alone. It is possible that further research may reveal complications which mask or interfere with an underlying process of segregation. Or it may be that segregation does not occur owing to some definite physiological reason which at present we do not understand.

And here it is impossible not to recall Mendel's own experiences with the Hawkweeds (*Hieracium*). This {133}genus of plants exhibits an extraordinary profusion of forms differing from one another sometimes in a single feature, sometimes in several. The question as to how far these numerous forms were to be classified as distinct species, how far as varieties, and how far as products of chance hybridisation, was even at that time a source of keen controversy among botanists. There is little doubt that Mendel undertook his experiments on the Hawkweeds in the hope that the conception of unit-characters so brilliantly demonstrated for the pea would serve to explain the great profusion of forms among the Hieraciums. Owing to the minute size of their florets, these plants offer very considerable technical difficulties in the way of cross fertilisation. By dint of great perseverance and labour, however, Mendel succeeded in obtaining a few crosses between different forms. These hybrids were reared and a further generation produced from them, and, no doubt somewhat to Mendel's chagrin, every one of them proved to breed true. There was a complete absence of that segregation of characters which he had shown to exist in peas and beans, and had probably looked forward with some confidence to finding in *Hieracium*. More than thirty years passed before the matter was cleared up. To-day we know that the peculiar behaviour of the hybrid Hieraciums is due to the fact that they normally produce seed by a peculiar process of parthenogenesis. It is possible to take an unopened flower and to shear off with a {134}razor all the male organs together with the stigmata through which the pollen reaches the ovules. The flower, nevertheless, sets perfectly good seed. But the cells from which the seeds develop are not of the same nature as the normal ovules of a plant. They are not gametes but retain the double structure of the maternal cells. They are rather to be regarded as of the nature of buds which early become detached from the parent stock to lead an independent existence, and, like buds, they reproduce exactly the maternal characteristics. The discovery of the true nature of this case was only rendered possible by the development of the study of cytology, and it was not given to Mendel to live long enough to learn why his hybrid Hieraciums all bred true.

{135}

CHAPTER XIII

VARIATION AND EVOLUTION

Through the facts of heredity we have reached a new conception of the individual. Hitherto we have been accustomed to distinguish between the members of a family of rabbits like that illustrated on Plate I. by assigning to each an individuality, and by making use of certain external features, such as the coat colour or the markings, as convenient outward signs to express our idea that the individuality of these different animals is different. Apart from this, our notions as to what constituted the individuality in each case were at best but vague. Mendelian analysis has placed in our hands a more precise method of estimating and expressing the variations that are to be found between one individual and another. Instead of looking at the individual as a whole, which is in some vague way endowed with an individuality marking it off from its fellows, we now regard it as an organism built up of definite characters superimposed on a basis beyond which for the moment our analysis will not take us. We have begun to realise that each individual has a definite architecture, and that this architecture depends {136}primarily upon the number and variety of the factors that existed in the two gametes that went to its building. Now most species exhibit considerable variation and exist in a number, often very large, of more or less well-defined varieties. How far can this great variety be explained in terms of a comparatively small number of factors if the number of possible forms depends upon the number of the factors which may be present or absent?

In the simple case where the homozygous and heterozygous conditions are indistinguishable in appearance the number of possible forms is 2, raised to the power of the number of factors concerned. Thus where one factor is concerned there are only $2^1 = 2$ possible forms, where ten factors are concerned there are $2^{10} = 1024$ possible forms differing from one another in at most ten and at least one character. Where the factors interact upon one another this number will, of course, be considerably increased. If the heterozygous form is different in appearance from the homozygous form, there are three possible forms connected with each factor; for ten such factors the possible number of individuals would be $3^{10} = 59,049$; for twenty such factors the possible number of different individuals would be $3^{20} = 3,486,784,401$. The presence or absence of a comparatively small number of factors in a species carries with it the possibility of an enormous range of individual variation. But every one of these individuals has a perfectly definite constitution which can {137}be determined in each case by the ordinary methods of Mendelian analysis. For in every instance the variation depends upon the presence or absence of definite factors carried in by the gametes from whose union the individual results. And as these factors separate out cleanly in the gametes which the individual forms, such variations as depend upon them are transmitted strictly according to the Mendelian scheme. Provided that the constitution of the gametes is unchanged, the heredity of such variation is independent of any change in the conditions of nutrition or environment which may operate upon the individual producing the gametes.

But, as everybody knows, an individual organism, whether plant or animal, reacts, and often reacts markedly, to the environmental conditions under which its life is passed. More especially is this to be seen where such characters as size or weight are concerned. More sunlight or a richer soil may mean stronger growth in a plant, better nutrition may result in a finer animal, superior education may lead to a more intelligent man. But although the changed conditions produce a direct effect upon the individual, we have no indisputable evidence that such alterations are connected with alterations in the nature of the gametes which the individual produces. And without this such variations cannot be perpetuated through heredity, but the conditions which produce the effect must always be renewed in each {138}successive generation. We are led, therefore, to the conclusion that two sorts of variations exist, those which are due to the presence of specific factors in the organism and those which are due to the direct effect of the environment during its lifetime. The former are known as *mutations*, and are inherited according to the Mendelian scheme; the latter have been termed *fluctuations*, and at present we have no valid reason for supposing that they are ever inherited. For though instances may be found in which effects produced during the lifetime of the individual would appear to affect the offspring, this is not necessarily due to heredity. Thus plants which are poorly nourished and grown under adverse conditions may set seed from which come plants that are smaller than the normal although grown under most favorable conditions. It is natural to attribute the smaller size of the offspring to the conditions under which the parents were grown, and there is no doubt that we should be quite right in doing so. Nevertheless, it need have nothing to do with heredity. As we have already pointed out, the seed is a larval plant which draws its nourishment from the mother. The size of the offspring is affected because the poorly nourished parent offered a bad environment to the young plant, and not because the gametes of the parent were changed through the adverse conditions under which it grew. The parent in this case is not only the producer of gametes, but also a part of the environment of the young {139}plant, and it is in this latter capacity that it affects its offspring. Wherever, as in plants and mammals, the organism is parasitic upon the mother during its earlier stages, the state of nutrition of the latter will almost certainly react upon it, and in this way a semblance of transmitted weakness or vigour is brought about. Such a connection between mother and offspring is purely one of environment, and it cannot be too strongly emphasised that it has nothing to do with the ordinary process of heredity.

The distinction between these two kinds of variation, so entirely different in their causation, renders it possible to obtain a clearer view of the process of evolution than that recently prevalent. As Darwin long ago realised, any theory of evolution must be based upon the facts of heredity and variation. Evolution only comes about through the survival of certain variations and the elimination of others. But to be of any moment in evolutionary change a variation must be inherited. And to be inherited it must be represented in the gametes. This, as we have seen, is the case for those variations which we have termed mutations. For the inheritance of fluctuations, on the other hand, of the variations which result from the direct action of the environment upon the individual, there is no indisputable evidence. Consequently we have no reason for regarding them as playing any part in the production of that succession of temporarily stable forms which we term evolution. In {140}the light of our present knowledge we must regard the mutation as the basis of evolution — as the material upon which natural selection works.

For it is the only form of variation of whose heredity we have any certain knowledge.

It is evident that this view of the process of evolution is in some respects at variance with that generally held during the past half century. There we were given the conception of an abstract type representing the species, and from it most of the individuals diverged in various directions, though, generally speaking, only to a very small extent. It was assumed that any variation, however small, might have a selection value, that is to say, could be transmitted to the offspring. Some of these would possess it in a less and some in a greater degree than the parent. If the variation were a useful one, those possessing to a rather greater extent would be favoured through the action of natural selection at the expense of their less fortunate brethren, and would leave a greater number of offspring, of whom some possessed it in an even more marked degree than themselves. And so it would go on. The process was a cumulative one. The slightest variation in a favourable direction gave natural selection a starting-point to work on. Through the continued action of natural selection on each successive generation the useful variation was gradually worked up, until at last it reached the magnitude of a specific {141}distinction. Were it possible in such a case to have all the forms before us, they would present the appearance of a long series imperceptibly grading from one extreme to the other.

Upon this view are made two assumptions not unnatural in the absence of any exact knowledge of the nature of heredity and variation. It was assumed, in the first place that variation was a continuous process, and, second, that any variation could be transmitted to the offspring. Both of these assumptions have since been shown to be unjustified. Even before Mendel's work became known Bateson had begun to call attention to the prevalence of discontinuity in variation, and a few years later this was emphasised by the Dutch botanist Hugo de Vries in his great work on *The Mutation Theory*. The ferment of new ideas was already working in the solution, and under the stimulus of Mendel's work they have rapidly crystallised out. With the advent of heredity as a definite science we have been led to revise our views as to the nature of variation, and consequently in some respects as to the trend of evolution. Heritable variation has a definite basis in the gamete, and it is to the gamete, therefore, not to the individual, that we must look for the initiation of this process. Somewhere or other in the course of their production is added or removed the factor upon whose removal or addition the new variation owes its existence. The new variation springs into being by a {142}sudden step, not by a process of gradual and almost imperceptible augmentation. It is not continuous but discontinuous, because it is based upon the presence or absence of some definite factor or factors—upon discontinuity in the gametes from which it sprang. Once formed, its continued existence is subject to the arbitrament of natural selection. If of value in the struggle for existence natural selection will decide that those who possess it shall have a better chance of survival and of leaving offspring than those who do not possess it. If it is harmful to the individual natural selection will soon bring about its elimination. But if the new variation is neither harmful nor useful there seems no reason why it should not persist.

In this way we avoid a difficulty that beset the older view. For on that view no new character could be developed except by the piling up of minute variations through the action of natural selection. Consequently any character found in animals and plants must be supposed to be of some definite use to the individual. Otherwise it could not have developed through the action of natural selection. But there are plenty of characters to which it is exceedingly difficult to ascribe any utility, and the ingenuity of the supporters of this view has often been severely taxed to account for their existence. On the more modern view this difficulty is avoided. The origin of a new variation is independent of natural {143}selection, and provided that it is not directly harmful there is no reason why it should not persist. In this way we are released from the burden of discovering a utilitarian motive behind all the multitudinous characters of living organisms. For we now recognise that the function of natural selection is selection and not creation. It has nothing to do with the formation of the new variation. It merely decides whether it is to survive or to be eliminated.

One of the arguments made use of by supporters of the older view is that drawn from the study of adaptation. Animals and plants are as a rule remarkably well adapted to living the life which their surroundings impose upon them, and in some cases this adaptation is exceedingly striking. Especially is this so in the many instances of what is called protective coloration, where the animal comes to resemble its surroundings so closely that it may reasonably be supposed to cheat even the keenest sighted enemy. Surely, we are told, such perfect adaptation could hardly have arisen through the mere survival of chance sports. Surely there must be some guiding hand moulding the species into the required shape. The argument is an old one. For John Ray that guiding hand was the superior wisdom of the Creator: for the modern Darwinian it is Natural Selection controlling the direction of variation. Mendelism certainly offers no suggestion of any such controlling force. It interprets the {144}variations of living forms in terms of definite physiological factors, and the diversity of animal and plant life is due to the gain or loss of these factors, to the origination of new ones, or to fresh combinations among those already in existence. Nor is there any valid reason against the supposition that even the most remarkable cases of resemblance, such as that of the leaf insect, may have arisen through a process of mutation. Experience with domestic plants and animals shows that the most bizarre forms may arise as sports and perpetuate themselves. Were such forms, arising under natural conditions, to be favoured by natural selection owing to a resemblance to something in their environment we should obtain a striking case of protective adaptation. And here it must not be forgotten that those striking cases to which our attention is generally called are but a very small minority of the existing forms of life.

For that special group of adaptation phenomena classed under the head of Mimicry, Mendelism seems to offer an interpretation simpler than that at present in vogue. This perhaps may be more clearly expressed by taking a specific case. There is in Africa a genus of Danaine butterflies known as *Amauris*, and there are reasons for considering that the group to which it belongs possesses properties which render it unpalatable to vertebrate enemies such as birds or monkeys. In the same region is also found the genus *Euralia* belonging to the entirely {145}different family of the Nymphalidae, to which there is no evidence for assigning the disagreeable properties of the Danaines. Now the different species of *Euralia* show remarkably close resemblances to the species of *Amauris*, which are found flying in the same region, and it is supposed that by "mimicking" the unpalatable forms they impose upon their enemies and thereby acquire immunity from attack. The point at issue is the way in which this seemingly purposeful resemblance has been brought about.

One of the species of *Euralia* occurs in two very distinct forms (Pl. VI.), which were previously regarded as separate species under the names *E. wahlbergi* and *E. mima*. These two forms respectively resemble *Amauris dominicanus* and *A. echeria*. For purposes of argument we will assume *A. echeria* to be the more recent form of the two. On the modern Darwinian view certain individuals of *A. dominicanus* gradually diverged from the *dominicanus* type and eventually reached the *echeria* type, though why this should have happened does not appear to be clear. At the same time those specimens which tended to vary in the direction of *A. echeria* in places where this species was more abundant than *A. dominicanus* were encouraged by natural selection, and under its guiding hand the form *mima* eventually arose from *wahlbergi*.

According to Mendelian views, on the other hand, {146} *A. echeria* arose suddenly from *A. dominicanus* (or *vice versa*), and similarly *mima* arose suddenly from *wahlbergi*. If *mima* occurred where *A. echeria* was common and *A. dominicanus* was rare, its resemblance to the more plentiful distasteful form would give it the advantage over *wahlbergi* and allow it to establish itself in place of the latter. On the modern Darwinian view natural selection gradually shapes *wahlbergi* into the *mima* form owing to the presence of *A. echeria*; on the Mendelian view natural selection merely conserves the *mima* form when once it has arisen. Now this case of mimicry is one of especial interest, because we have experimental evidence that the relation between *mima* and *wahlbergi* is a simple Mendelian one, though at present it is uncertain which is the dominant and which the recessive form. The two have been proved to occur in families bred from the same female without the occurrence of any intermediates, and the fact that the two segregate cleanly is strong evidence in favour of the Mendelian view. On this view the genera *Amauris* and *Euralia* contain a similar set of pattern factors, and the conditions, whatever they may be, which bring about mutation in the former lead to the production of a similar mutation in the latter. Of the different forms of *Euralia* produced in any region that one has the best chance of survival, through the operation of natural selection, which resembles the most plentiful *Amauris* form. Mimetic resemblance is a true phenomenon, but natural selection plays the part of a conservative, not of a formative agent.

Plate VI.

{147}

It is interesting to recall that in earlier years Darwin was inclined to ascribe more importance to "sports" as opposed to continuous minute variation, and to consider that they might play a not inconsiderable part in the formation of new varieties in nature. This view, however, he gave up later, because he thought that the relatively rare sport or mutation would rapidly disappear through the swamping effects of crossing with the more abundant normal form, and so, even though favoured by natural selection, would never succeed in establishing itself. Mendel's discovery has eliminated this difficulty. For suppose that the sport differed from the normal in the loss of a factor and were recessive. When mated with the normal this character would seem to disappear, though, of course, half of the gametes of its progeny would bear it. By continual crossing with normals a small proportion of heterozygotes would eventually be scattered among the population, and as soon as any two of these mated together the recessive sport would appear in one quarter of their offspring.

A suggestive contribution to this subject was recently made by G. H. Hardy. Considering the distribution of a single factor in a mixed population consisting of the heterozygous and the two homozygous forms he showed that such a population breeding at random rapidly fell into a {148}stable condition with regard to the proportion of these three forms, whatever may have been the proportion of the three forms to start with. Let us suppose for instance, that the population consists of p homozygotes of one kind, r homozygotes of the other kind, and $2q$ heterozygotes. Hardy pointed out that, other things being equal, such a population would be in equilibrium for this particular factor so long as the condition $q^2 = pr$ was fulfilled. If the condition is fulfilled to start with, the population remains in equilibrium. If the condition is not fulfilled to start with, Hardy showed that a position of equilibrium becomes established after a single generation, and that this position is thereafter maintained. The proportions of the three classes which satisfy the equation $q^2 = pr$ are exceedingly numerous, and populations in which they existed in the proportions shown in the appended table would remain in stable equilibrium generation after generation:—

$p.$	$2q.$	$r.$
1	2	1
1	4	4
1	6	9
1	8	16
1	20,000	100,000,000
1	$2n$	n^2

This, of course, assumes that all three classes are equally fertile, and that no form of selection is taking place to the {149}benefit of one class more than of another. Moreover, it makes no difference whether p represents the homozygous dominants or whether it stands for the recessives. A population containing a very small proportion of dominants and one containing a similar proportion of recessives are equally stable. The term dominant is in some respects apt to be misleading, for a dominant character cannot in virtue of its dominance establish itself at the expense of a recessive one. Brown eyes in man are dominant to blue, but there is no reason to suppose that as years go on the population of these islands will become increasingly brown eyed. Given equality of conditions both are on an equal footing. If, however, either dominant or recessive be favoured by selection the conditions are altered, and it can be shown that even a small advantage possessed by the one will rapidly lead to the elimination of the other. Even with but a 5 per cent selection advantage in its favour it can be shown that a rare sport will oust the normal form in a few hundred generations. In this way we are freed from a difficulty inherent in the older view that varieties arose through a long-continued process involving the accumulation of very slight variations. On that view the establishing of a new type was of necessity a very long and tedious business, involving many thousands of generations. For this reason the biologist has been accustomed to demand a very large supply of time, often a great deal more than the physicist is {150}disposed to grant, and this has sometimes led him to expostulate with the latter for cutting off the supply. On the newer views, however, this difficulty need not arise, for we realise that the origin and establishing of a new form may be a very much more rapid process than has hitherto been deemed possible.

One last question with regard to evolution. How far does Mendelism help us in connection with the problem of the origin of species? Among the plants and animals with which we have dealt we have been able to show that distinct differences, often considerable, in colour, size, and structure, may be interpreted in terms of Mendelian factors. It is not unlikely that most of the various characters which the systematist uses to mark off one species from another, the so-called specific characters, are of this nature. They serve as convenient labels, but are not essential to the conception of species. A systematist who defined the wild sweet pea could hardly fail to include in his definition such characters as the procumbent habit, the tendrils, the form of the pollen, the shape of the flower, and its purple colour. Yet all these and other characters have been proved to depend upon the presence of definite factors which can be removed by appropriate crossing. By this means we can produce a small plant a few inches in height with an erect habit of growth, without tendrils, with round instead of oblong pollen, and with colourless deformed flowers quite different {151}in appearance from those of the wild form. Such a plant would breed perfectly true, and a botanist to whom it was presented, if ignorant of its origin, might easily relegate it to a different genus. Nevertheless, though so widely divergent in structure, such a plant must yet be regarded as belonging to the species *Lathyrus odoratus*. For it still remains fertile with the many different varieties of sweet pea. It is not visible attributes that constitute the essential difference between one species and another. The essential difference, whatever it may be, is that underlying the phenomenon of sterility. The visible attributes are those made use of by the systematist in cataloguing the different forms of animal and plant life, for he has no other choice. But it must not be forgotten that they are often misleading. Until they were bred together *Euralia wahlbergi* and *E. mima* were regarded as perfectly valid species, and there is little doubt that numbers of recognised species will eventually fall to the ground in the same way as soon as we are in a position to apply the test of breeding. Mendelism has helped us to realise that specific characters may be but incidental to a species—that the true criterion of what constitutes a species is sterility, and that particular form of sterility which prevents two healthy gametes on uniting from producing a zygote with normal powers of growth and reproduction. For there are forms of sterility which are purely mechanical. The pollen of *Mirabilis julupu* cannot fertilise *M.* {152} *longiflora*, because the pollen tubes of the former are not long enough to penetrate down to the ovules of the latter. Hybrids can nevertheless be obtained from the reciprocal cross. Nor should we expect offspring from a St. Bernard and a toy terrier without recourse to artificial fertilisation. Or sterility may be due to pathological causes which prevent the gametes from meeting one another in a healthy state. But in most cases it is probable that the sterility is due to some other cause. It is not inconceivable that definite differences in chemical composition render the protoplasm of one species toxic to the gametes of the other, and if this is so it is not impossible that we may some day be able to express these differences in terms of Mendelian factors. The very nature of the case makes it one of extreme difficulty for experimental investigation. At any rate, we realise more clearly than before that the problem of species is not one that can be resolved by the study of morphology or of systematics. It is a problem in physiology.

{153}

CHAPTER XIV

ECONOMICAL

Since heredity lies at the basis of the breeder's work, it is evident that any contribution to a more exact knowledge of this subject must prove of service to him, and there is no doubt that he will be able to profit by Mendelian knowledge in the conduct of his operations. Indeed, as we shall see later, these ideas have already led to striking results in the raising of new and more profitable varieties. In the first place, heredity is a question of individuals. Identity of appearance is no sure guide to reproductive qualities. Two individuals similarly bred and indistinguishable in outward form may nevertheless behave entirely differently when bred from. Take, for instance, the family of sweet peas shown on Plate IV. The F_2 generation here consists of seven distinct types, three sorts of purples, three sorts of reds, and whites. Let us suppose that our object is to obtain a true breeding strain of the pale purple picotee form. Now from the proportions in which they come we know that the dilute colour is due to the absence of the factor which intensifies the colour. Consequently the picotee cannot throw the {154}two deeper shades of red or purple. But it may be heterozygous for the purpling factor when it will throw the dilute red (tinged white), or it may be heterozygous for either or both of the two colour factors (cf. p. 44), in which case it will throw whites. Of the picotees which come in such a family, therefore, some will give picotees, tinged whites, and whites, others will give picotees and tinged whites only, others will give picotees and whites only, while others, again, and these the least numerous, will give nothing but picotees. The new variety is already fixed in a certain definite proportion of the plants; in this particular instance in 1 out of every 27. All that remains to be done is to pick out these plants. Since all the picotees look alike, whatever their breeding capacity, the only way to do this is to save the seed from a number of such plants *individually*, and to raise a further generation. Some of them will be found to breed true. The variety is then established, and may at once be put on the market with full confidence that it will hereafter throw none of the other forms. The all-important thing is to save and sow the seed of separate individuals separately. However alike they look, the seed from different individuals must on no account be mixed. Provided that due care is taken in this respect no long and tedious process of selection is required for the fixation of any given variety. Every possible variety arising from a cross appears in the F_2 generation if only a sufficient {155}number is raised, and of all these different varieties a certain proportion of each is already fixed. Heredity is a question of individuals, and the recognition of this will save the breeder much labour, and enable him to fix his varieties in the shortest possible time.

Such cases as these of the sweet pea throw a fresh light upon another of the breeder's conceptions, that of purity of type. Hitherto the criterion of a "pure-bred" thing, whether plant or animal, has been its pedigree, and the individual was regarded as more or less pure bred for a given quality according as it could show a longer or shorter list of ancestors possessing this quality. To-day we realise that this is not essential. The pure-bred picotee appears in our F_2 family though its parent was a purple bicolor, and its remoter ancestors whites for generations. So also from the cross between pure strains of black and albino rabbits we may obtain in the F_2 generation animals of the wild agouti colour which breed as true to type as the pure wild rabbit of irreproachable pedigree. The true test of the pure breeding thing lies not in its ancestry but in the nature of the gametes which have gone to its making. Whenever two similarly constituted gametes unite, whatever the nature of the parents from which they arose, the resulting individual is homozygous in all respects and must consequently breed true. In deciding questions of purity it is to the gamete, and not to ancestry, that our appeal must henceforth be made. {156}

Improvement is after all the keynote to the breeder's operations. He is aiming at the production of a strain which shall combine the greatest number of desirable properties with the least number of undesirable ones. This good quality he must take from one strain, that from another, and that again from a third, while at the same time avoiding all the poor qualities that these different strains possess. It is evident that the Mendelian conception of characters based upon definite factors which are transmitted on a definite scheme must prove of the greatest service to him. For once these factors have been determined, their distribution is brought under control, and they can be associated together or dissociated at the breeder's will. The chief labour involved is that necessary for the determination of the factors upon which the various characters depend. For it often happens that what appears to be a simple character turns out when analysed to depend upon the simultaneous presence of several distinct factors. Thus the Malay fowl breeds true to the walnut comb, as does also the Leghorn to the single comb, and when pure strains are crossed all the offspring have walnut combs. At first sight it would be not unnatural to regard the difference as dependent upon the presence or absence of a single factor. Yet, as we have already seen, two other types of comb, the pea and the rose, make their appearance in the F_2 generation. Analysis shows that the difference between the walnut {157}and the single is a difference of two factors, and it is not until this has been determined that we can proceed with certainty to transfer the walnut character to a single-combed breed. Moreover, in his process of analysis the breeder must be prepared to encounter the various phenomena that we have described under the headings of interaction of factors, coupling and repulsion, and the recognition of these phenomena will naturally influence his procedure. Or again, his experiments may show him that one of the characters he wants, like the blue of the Andalusian fowl, is dependent upon the heterozygous nature of the individual which exhibits it, and if such is the case he will be wise to refrain from any futile attempt at fixing it. If it is essential it must be built up again in each generation, and he will recognise that the most economical way of doing this is to cross the two pure strains so that all the offspring may possess the desired character. The labour of analysis is often an intricate and tedious business. But once done it is done once for all. As soon as the various factors are determined, upon which the various characters of the individual depend, as soon as the material to be made use of has been properly analysed, the production and fixation of the required combinations becomes a matter of simple detail.

An excellent example of the practical application of Mendelian principles is afforded by the experiments which Professor Biffen has recently carried out in Cambridge. {158}Taken as a whole English wheats compare favourably with foreign ones in respect of their cropping power. On the other hand, they have two serious defects.

They are liable to suffer from the attacks of the fungus which causes rust, and they do not bake into a good loaf. This last property depends upon the amount of gluten present, and it is the greater proportion of this which gives to the "hard" foreign wheat its quality of causing the loaf to rise well when baked. For some time it was held that "hard" wheat with a high glutinous content could not be grown in the English climate, and undoubtedly most of the hard varieties imported for trial deteriorated greatly in a very short time. Professor Biffen managed to obtain a hard wheat which kept its qualities when grown in England. But in spite of the superior quality of its grain from the baker's point of view its cropping capacity was too low for it to be grown profitably in competition with English wheats. Like the latter, it was also subject to rust. Among the many varieties which Professor Biffen collected and grew for observation he managed to find one which was completely immune to the attacks of the rust fungus, though in other respects it had no desirable quality to recommend it. Now as the result of an elaborate series of investigations he was able to show that the qualities of heavy cropping capacity, "hardness" of grain, and immunity to rust can all be expressed in terms of Mendelian factors. Having once analysed his material {159}the rest was comparatively simple, and in a few years he has been able to build up a strain of wheat which combines the cropping capacity of the best English varieties with the hardness of the foreign kinds, and at the same time is completely immune to rust. This wheat has already been shown to keep its qualities unchanged for several years, and there is little doubt that when it comes to be grown in quantity it will exert an appreciable influence on wheat-growing in Great Britain.

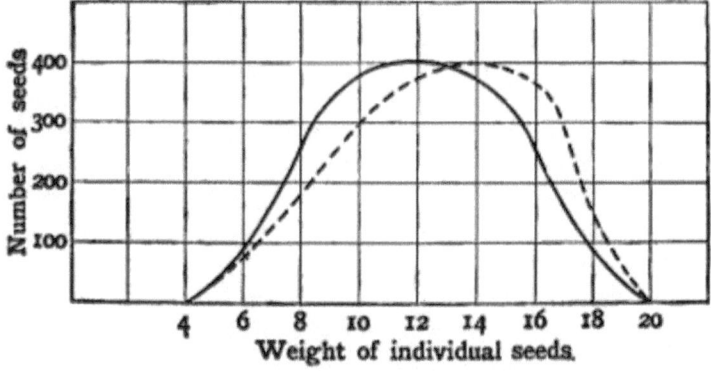

Fig. 30.

Curves to illustrate the influence of selection.

It may be objected that it is often with small differences rather than with the larger and more striking ones that the breeder is mainly concerned. It does not matter much to him whether the colour of a pea flower is purple or pink or white. But it does matter whether the plant bears rather larger seeds than usual, or rather more of them. Even a small difference when multiplied by the {160}size of the crop will effect a considerable difference in the profit. It is the general experience of seedsmen and others that differences of this nature are often capable of being developed up to a certain point by a process of careful selection each generation. At first sight this appears to be something very like the gradual accumulation of minute variations through the continuous application of a selective process. Some recent experiments by Professor Johannsen of Copenhagen set the matter in a different light. One of his investigations deals with the inheritance of the weight of beans, but as an account of these experiments would involve us in the consideration of a large amount of detail we may take a simple imaginary case to illustrate the nature of the conclusions at which he arrived. If we weigh a number of seeds collected from a patch of plants such as Johannsen's beans we should find that they varied considerably in size. The majority would probably not diverge very greatly from the general average, and as we approached the high or low extreme we should find a constantly decreasing number of individuals with these weights. Let us suppose that the weight of our seed varied between 4 and 20 grains, that the greatest number of seeds were of the mean weight, viz. 12 grains, and that as we passed to either extreme at 4 and 20 the number became regularly less. The weight relation of such a collection of seeds can be expressed by the accompanying curve (Fig. 30). Now if we select for {161}sowing only that seed which weighs over 12 grains, we shall find that in the next generation the average weight of the seed is raised and the curve becomes somewhat shifted to the right as in the dotted line of Fig. 30. By continually selecting we can shift our curve a little more to the right, *i.e.* we can increase the average weight of the seeds until at last we come to a limit beyond which further selection has no effect. This phenomenon has been long known, and it was customary to regard these variations as of a continuous nature, *i.e.* as all chance fluctuations in a homogeneous mass, and the effect of selection was supposed to afford evidence that small continuous variations could be increased by this process. But Johannsen's results point to another interpretation. Instead of our material being homogeneous it is probably a mixture of several strains each with its own average weight about {162}which the varying conditions of the environment cause it to fluctuate. Each of these strains is termed a **pure line**. If we imagine that there are three such pure lines in our imaginary case, with average weights 10, 12, 14 grains respectively, and if the range of fluctuation of each of these pure lines is 12 grains, then our curve must be represented as made up of the three components

A	fluctuating	between	4	and	16	with	a mean of	10
B	"	"	6	"	18	"	"	12
C	"	"	8	"	20	"	"	14

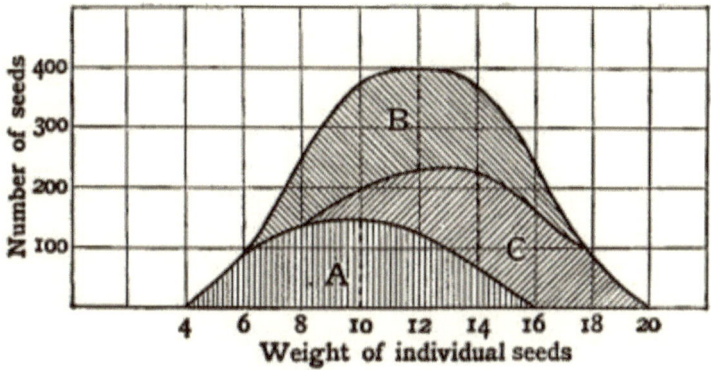

Fig. 31.

Curves to illustrate the conception of pure lines in a population.

as is shown in Fig. 31. A seed that weighs 12 grains may belong to any of these three strains. It may be an average seed of B, or a rather large seed of A, or a rather small seed of C. If it belongs to B its offspring will average 12 grains, if to A they will average 10 grains, and if to C they will average 14 grains. Seeds of similar weight may give a different result because they happen to be fluctuations of different pure lines. But within the pure line any seed, large or small, produces the average result for that line. Thus a seed of line C which weighs 20 grains will give practically the same result as one that weighs 10 grains.

On this view we can understand why selection of the largest seed raises the average weight in the next generation. We are picking out more of C and less of A and B, and as this process is repeated the proportion of C gradually increases and we get the appearance of selection {163}acting on a continuously varying homogeneous material and producing a permanent effect. This is because the interval between the average weight of the different pure lines is small compared with the environmental fluctuations. None the less it is there, and the secret of separating and fixing any of these pure lines is again to breed from the individual separately. As soon as the pure line is separated further selection becomes superfluous.

Since the publication of Darwin's famous work upon the effects of cross and self fertilisation, it has been generally accepted that the effect of a cross is commonly, though not always, to introduce fresh vigour into the offspring, though why this should be so we are quite at a loss to explain. Continued close inbreeding, on the contrary, eventually leads to deterioration, though, as in many self-fertilised plants, a considerable number of generations may elapse before it shows itself in any marked degree. The fine quality of many of the seedsman's choice varieties of vegetables probably depends upon the fact that they had resulted from a cross but a few generations back, and it is possible that they often oust the older kinds not because they started as something intrinsically better, but because the latter had gradually deteriorated through continuous self-fertilisation. Most breeders are fully alive to the beneficial results of a cross so far as vigour is concerned, but they often hesitate to embark upon it owing to what was held {164}to be the inevitably lengthy and laborious business of recovering the original variety and refixing it, even if in the process it was not altogether lost. That danger Mendelism has removed, and we now know that by working on these lines it is possible in three or four generations to recover the original variety in a fixed state with all the superadded vigour that follows from a cross.

Nor is the problem one that concerns self-fertilised plants only. Plants that are reproduced asexually often appear to deteriorate after a few generations unless a sexual generation is introduced. New varieties of potato, for example, are frequently put upon the market, and their excellent qualities give them a considerable vogue. Much is expected of them, but time after time they deteriorate in a disappointing way and are lost to sight. It is not improbable that we are here concerned with a case in which the plants lose their vigour after a few asexual generations of reproduction from tubers, and can only recover it with the stimulus that results from the interpolation of a sexual generation. Unfortunately this generally means that the variety is lost, for owing to the haphazard way in which new kinds of potatoes are reproduced it is probable that most cultivated varieties are complex heterozygotes. Were the potato plant subjected to careful analysis and the various factors determined upon which its variations depend, we should be in a position to remake continually any good potato without {165}running the risk of losing it altogether, as is now so often the case.

The application of Mendelian principles is likely to prove of more immediate service for plants than animals, for owing to the large numbers which can be rapidly raised from a single individual and the prevalence of self-fertilisation, the process of analysis is greatly simplified. Even apart from the circumstance that the two sexes may sometimes differ in their powers of transmission, the mere fact of their separation renders the analysis of their properties more difficult. And as the constitution of the individual is determined by the nature and quality of its offspring, it is not easy to obtain this knowledge where the offspring, as in most animals, are relatively few. Still, as has been abundantly shown, the same principles hold good here also, and there is no reason why the process of analysis, though more troublesome, should not be effectively carried out. At the same time, it affords the breeder a rational basis for some familiar but puzzling phenomena. The fact, for instance, that certain characters often "skip a generation" is simply the effect of dominance in F_1 and the reappearance of the recessive character in the following generation. "Reversion" and "atavism," again, are phenomena which are no longer mysterious, but can be simply expressed in Mendelian terms as we have already suggested in Chap. VI. The occasional appearance of a sport in a supposedly pure strain is {166}often due to the reappearance of a recessive character. Thus even in the most highly pedigreed strains of polled cattle such as the Aberdeen Angus, occasional individuals with horns appear. The polled character is dominant to the horned, and the occasional reappearance of the horned animal is due to the fact that some of the polled herd are heterozygous in this character. When two such individuals are mated, the chances are 1 in 4 that the offspring will be horned. Though the heterozygous individuals may be indistinguishable in appearance from the pure dominant, they can be readily separated by the breeding test. For when crossed by the recessive, in this case horned animals, the pure dominant gives only polled beasts, while the heterozygous individual gives equal numbers of polled and horned ones. In this particular instance it would probably be impracticable to test all the cows by crossing with a horned bull. For in each case it would be necessary to have several polled calves from each before they could with reasonable certainty be regarded as pure dominants. But to ensure that no horned calves should come, it is enough to use a bull which is pure for that character. This can easily be tested by crossing him with a dozen or so horned cows. If he gets no horned calves out of these he may be regarded as a pure dominant and thenceforward put to his own cows, whether horned or polled, with the certainty that all his calves will be polled. {167}

Or, again, suppose that a breeder has a chestnut mare and wishes to make certain of a bay foal from her. We know that bay is dominant to chestnut, and that if a homozygous bay stallion is used a bay foal must result. In his choice of a sire, therefore, the breeder must be guided by the previous record of the animal, and select one that has never given anything but bays when put to either bay or chestnut mares. In this way he will assure himself of a bay foal from his chestnut mare, whereas if the record of the sire shows that he has given chestnuts he will be heterozygous, and the chances of his getting a bay or a chestnut out of a chestnut mare are equal.

It is not impossible that the breeder may be unwilling to test his animals by crossing them with a different breed through fear that their purity may be thereby impaired, and that the influence of the previous cross may show itself in succeeding generations. He might hesitate, for instance, to test his polled cows by crossing them with a horned bull for fear of getting horned calves when the cows were afterwards put to a polled bull of their own breed. The belief in the power of a sire to influence subsequent generations, or telegony as it is sometimes called, is not uncommon even to-day. Nevertheless, carefully conducted experiments by more than one competent observer have failed to elicit a single shred of unequivocal evidence in favour of the view. Until we have evidence based upon experiments which are capable of {168}repetition, we may safely ignore telegony as a factor in heredity.

Heterozygous forms play a greater part in the breeding of animals than of plants, for many of the qualities sought after by the breeder are of this nature. Such is the blue of the Andalusian fowl, and, according to Professor Wilson, the roan of the Shorthorn is similar, being the heterozygous form produced by mating red with white. The characters of certain breeds of canaries and pigeons again appear to depend upon their heterozygous nature. Such forms cannot, of course, ever be bred true, and where several factors are concerned they may when bred together produce but a small proportion of offspring like themselves. As soon, however, as their constitution has been analysed and expressed in terms of Mendelian factors, pure strains can be built up which when crossed will give nothing but offspring of the desired heterozygous form.

The points with which the breeder is concerned are often fine ones, not very evident except to the practised eye. Between an ordinary Dutch rabbit and a winner, or between the comb of a Hamburgh that is fit to show and one that is not, the differences are not very apparent to the uninitiated. Whether Mendelism will assist the breeder in the production of these finer points is at present doubtful. It may be that these small differences are heritable, such as those that form the basis of Johannsen's pure lines. In this case the breeder's outlook is {169}hopeful. But it may be that the variations which he seeks to perpetuate are of the nature of fluctuations, dependent upon the earlier life conditions of the individual, and not upon the constitution of the gametes by which it was formed. If such is the case, he will get no help from the science of heredity, for we know of no evidence which might lead us to suppose that variations of this sort can ever become fixed and heritable.

{170}

CHAPTER XV

MAN

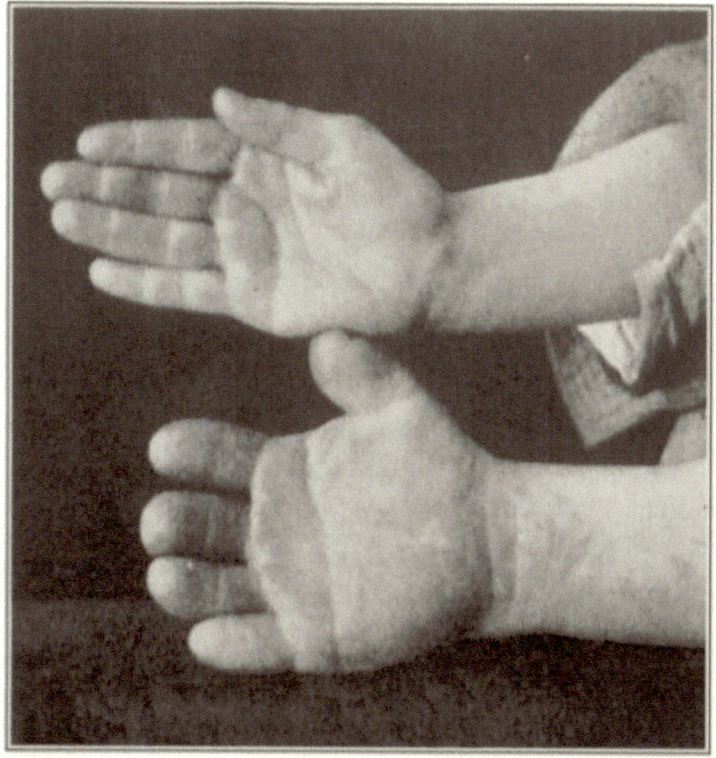

Fig. 32.

Normal and brachydactylous hands placed together for comparison. (From Drinkwater.)

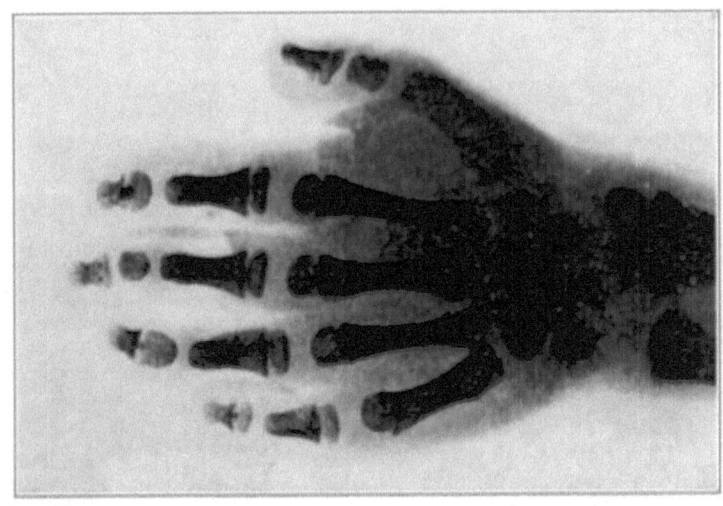

Fig. 33.

Radiograph of a brachydactylous hand.

Though the interest attaching to heredity in man is more widespread than in other animals, it is far more difficult to obtain evidence that is both complete and accurate. The species is one in which the differentiating characters separating individual from individual are very numerous, while the number of the offspring is comparatively few, and the generations are far between. For these reasons, even if it were possible, direct experimental work with man would be likely to prove both tedious and expensive. There is, however, another method besides the direct one from which something can be learned. This consists in collecting all the evidence possible, arranging it in the form of pedigrees, and comparing it with standard cases already worked out in animals and plants. In this way it has been possible to demonstrate in man the existence of several characters showing simple Mendelian inheritance. As few besides medical men have hitherto been concerned practically with heredity, such records as exist are, for the most part, records of deformity or of disease. So it happens that most of the {171}pedigrees at present available deal with characters which are usually classed as abnormal. In some of these the inheritance is clearly Mendelian. One of the cases which has been most fully worked out is that of a deformity known as brachydactyly. In brachydactylous people the {172}whole of the body is much stunted, and the fingers and toes appear to have two joints only instead of three (cf. Figs. 32 and 33). The inheritance of this peculiarity has been carefully investigated by Dr. Drinkwater, who collected all the data he was able to find among the members of a large family in which it occurred. The result is the pedigree shown on p. 173. It is assumed that all who are recorded as having offspring were married to normals. Examination of the pedigree brings out the facts (1) that all affected individuals have an affected parent; (2) that none of the unaffected individuals, though sprung from the affected, ever have descendants who are affected, and (3) that in families where both affected and unaffected {173}occur, the numbers of the two classes are, on the average, equal. (The sum of such families in the complete pedigree is thirty-nine affected and thirty-six normals.) It is obvious that these are the conditions which are fulfilled in a simple Mendelian case, and there is nothing in this pedigree to contradict the assertion that brachydactyly, whatever it may be due to, behaves as a simple dominant to the normal form, *i.e.* that it depends upon a factor which the normal does not contain. The recessive normals cannot transmit the affected condition whatever their ancestry. Once free they are always free, and can marry other normals with full confidence that none of their children will show the deformity.

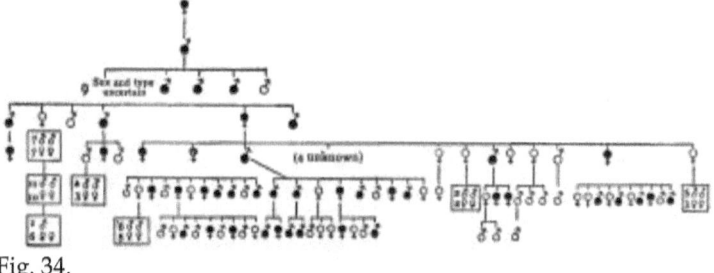

Fig. 34.

Pedigree of Drinkwater's brachydactylous family. The affected members are indicated by black and the normals by light circles.

{174}

The evidence available from pedigrees has revealed the simplest form of Mendelian inheritance in several human defects and diseases, among which may be mentioned presenile cataract of the eyes, an abnormal form of skin thickening in the palms of the hands and soles of the feet, known as tylosis, and epidermolysis bullosa, a disease in which the skin rises up into numerous bursting blisters.

Among the most interesting of all human pedigrees is one recently built up by Mr. Nettleship from the records of a night-blind family living near Monpelier in the south of France. In night-blind people the retina is insensitive to light which falls below a certain intensity, and such people are consequently blind in failing daylight or in moonlight. As the Monpelier case had excited interest for some time, the records are unusually complete. They commence with a certain Jean Nougaret, who was born in 1637, and suffered from night-blindness, and they end for the present with children who are to-day but a few years of age. Particulars are known of over 2000 of the descendants of Jean Nougaret. Through ten generations and nearly three centuries the affection has behaved as a Mendelian dominant, and there is no sign that long-continued marriage with folk of normal vision has produced any amelioration of the night-blind state. {175}

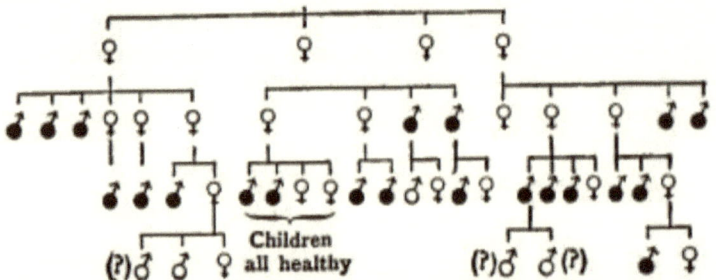

Fig. 35.

Pedigree of a hæmophilic family. Affected (all males) represented by black, and normals of both sexes by light circles. (From Stahel.)

Besides cases such as these where a simple form of Mendelian inheritance is obviously indicated, there are others which are more difficult to read. Of some it may be said that on the whole the peculiarity behaves as though it were an ordinary dominant; but that exceptions occur in which affected children are born to unaffected parents. It is not impossible that the condition may, like colour in the sweet pea, depend upon the presence or absence of more than one factor. In none of these cases, however, are the data sufficient for determining with certainty whether this is so or not.

A group of cases of exceptional interest is that in which the incidence of disease is largely, if not absolutely, restricted to one sex, and so far as is hitherto known the burden is invariably borne by the male. In the inheritance of colour-blindness (p. 117) we have already discussed an instance in which the defect is rare, though not {176}unknown, in the female. Sex-limited inheritance of a similar nature is known for one or two ocular defects, and for several diseases of the nervous system. In the peculiarly male disease known as hæmophilia the blood refuses to clot when shed, and there is nothing to prevent great loss from even a superficial scratch. In its general trend the inheritance of hæmophilia is not unlike that of horns among sheep, and it is possible that we are here again dealing with a character which is dominant in one sex and recessive in the other. But the evidence so far collected points to a difference somewhere, for in hæmophilic families the affected males, instead of being equal in number to the unaffected, show a considerable preponderance. The unfortunate nature of the defect, however, forces us to rely for our interpretation almost entirely upon the families produced by the unaffected females who can transmit it. Our knowledge of the offspring of "bleeding" males is as yet far too scanty, and until it is improved, or until we can find some parallel case in animals or plants, the precise scheme of inheritance for hæmophilia must remain undecided.

Though by far the greater part of the human evidence relates to abnormal or diseased conditions, a start has been made in obtaining pedigrees of normal characters. From the ease with which it can be observed, it was natural that eye-colour should be early selected as a subject of investigation, and the work of Hurst and others {177}has clearly demonstrated the existence of one Mendelian factor in operation here. Eyes are of many colours, and the colour depends upon the pigment in the iris. Some eyes have pigment on both sides of the iris—on the side that faces the retina as well as on the side that looks out upon the world. Other eyes have pigment on the retinal side only. To this class belong the blues and clear greys; while the eyes with pigment in front of the iris also are brown, hazel, or green in various shades according to the amount of pigment present. In albino animals the pigment is entirely absent, and as the little blood-vessels are not obscured the iris takes on its characteristic pinkish-red appearance. The condition in which pigment is present in front of the iris is dominant to that in which it is absent. Greens, browns, or hazels mated together may, if heterozygous, give the recessive blue, but no individuals of the brown class are to be looked for among the offspring of blues mated together. The blues, however, may carry factors which are capable of modifying the brown. Just as the pale pink-tinged sweet pea (Pl. IV., 9) when mated with a suitable white gives only deep purples, so an eye with very little brown pigment mated with certain blues produces progeny of a deep brown, far darker than either parent. The blue may carry a factor which brings about intensification of the brown pigment. There are doubtless other factors which modify the brown when present, but we do not yet know enough of the {178}inheritance of the various shades to justify any statement other than that the heredity of the pigment in front of the iris behaves as though it were due to a Mendelian factor.

Even this fact is of considerable importance, for it at once suggests that the present systems of classification of eye-colours, to which some anthropologists attach considerable weight, are founded on a purely empirical and unsatisfactory basis. Intensity of colour is the criterion at present in vogue, and it is customary to arrange the eye-colours in a scale of increasing depth of shade, starting with pale greys and ending with the deepest browns. On this system the lighter greens are placed among the blues. But we now know that blues may differ from the deep browns in the absence of only a single factor, while, on the other hand, the difference between a blue and a green may be a difference dependent upon more than one factor. To what extent eye-colour may be valuable as a criterion of race it is at present impossible to say, but if it is ever to become so, it will only be after a searching Mendelian analysis has disclosed the factors upon which the numerous varieties depend.

A discussion of eye-colour suggests reflections of another kind. It is difficult to believe that the markedly different states of pigmentation which occur in the same species are not associated with deep-seated chemical differences influencing the character and bent of the individual. {179}May not these differences in pigmentation be coupled with and so become in some measure a guide to mental and temperamental characteristics? In the National Portrait Gallery in London the pictures of celebrated men and women are largely grouped according to the vocations in which they have succeeded. The observant will probably have noticed that there is a tendency for a given type of eye-colour to predominate in some of the larger groups. It is rare to find anything but a blue among the soldiers and sailors, while among the actors, preachers, and orators the dark eye is predominant, although for the population as a whole it is far scarcer than the light. The facts are suggestive, and it is not impossible that future research may reveal an intimate connection between peculiarities of pigmentation and peculiarities of mind.

The inheritance of mental characters is often elusive, for it is frequently difficult to appraise the effects of early environment in determining a man's bent. That ability can be transmitted there is no doubt, for this is borne out by general experience, as well as by the numerous cases of able families brought together by Galton and others. But when we come to inquire more precisely what it is that is transmitted we are baffled. A distinguished son follows in the footsteps of a distinguished father. Is this due to the inheritance of a particular mental aptitude, or is it an instance of general mental ability displayed in a field rendered attractive by early association? We have {180}at present very little definite evidence for supposing that what appear to be special forms of ability may be due to specific factors. Hurst, indeed, has brought forward some facts which suggest that musical sense sometimes behaves as a recessive character, and it is likely that the study of some clean-cut faculty such as the mathematical one would yield interesting results.

The analysis of mental characters will no doubt be very difficult, and possibly the best line of attack is to search for cases where they are associated with some physical feature such as pigmentation. If an association of this kind be found, and the pigmentation factors be determined, it is evident that we should thereby obtain an insight into the nature of the units upon which mental conditions depend. Nor must it be forgotten that mental qualities, such as quickness, generosity, instability, etc.,—qualities which we are accustomed to regard as convenient units in classifying the different minds with which we are daily brought into contact, — are not necessarily qualities that correspond to heritable units. Effective mental ability is largely a matter of temperament, and this in turn is quite possibly dependent upon the various secretions produced by the different tissues of the body. Similar nervous systems associated with different livers might conceivably result in individuals upon whose mental ability the world would pass a very different judgment. Indeed, it is not at all impossible {181}that a particular form of mental ability may depend for its manifestation, not so much upon an essential difference in the structure of the nervous system, as upon the production by another tissue of some specific poison which causes the nervous system to react in a definite way. We have mentioned these possibilities merely to indicate how complex the problem may turn out to be. Though there is no doubt that mental ability is inherited, what it is that is transmitted, whether factors involving the quality and structure of the nervous system itself, or factors involving the production of specific poisons by other tissues, or both together, is at present uncertain.

Little as is known to-day of heredity in man, that little is of extraordinary significance. The qualities of men and women, physical and mental, depend primarily upon the inherent properties of the gametes which went to their making. Within limits these qualities are elastic, and can be modified to a greater or lesser extent by influences brought to bear upon the growing zygote, provided always that the necessary basis is present upon which these influences can work. If the mathematical faculty has been carried in by the gamete, the education of the zygote will enable him to make the most of it. But if the basis is not there, no amount of education can transform that zygote into a mathematician. This is a matter of common experience. Neither is there any reason for supposing that the superior education of a {182}mathematical zygote will thereby increase the mathematical propensities of the gametes which live within him. For the gamete recks little of quaternions. It is true that there is progress of a kind in the world, and that this progress is largely due to improvements in education and hygiene. The people of to-day are better fitted to cope with their material surroundings than were the people of even a few thousand years ago. And as time goes on they are able more and more to control the workings of the world around them. But there is no reason for supposing that this is because the effects of education are inherited. Man stores knowledge as a bee stores honey or a squirrel stores nuts. With man, however, the hoard is of a more lasting nature. Each generation in using it sifts, adds, and rejects, and passes it on to the next a little better and a little fuller. When we speak of progress we generally mean that the hoard has been improved, and is of more service to man in his attempts to control his surroundings. Sometimes this hoarded knowledge is spoken of as the inheritance which a generation receives from those who have gone before. This is misleading. The handing on of such knowledge has nothing more to do with heredity in the biological sense than has the handing on from parent to offspring of a picture, or a title, or a pair of boots. All these things are but the transfer from zygote to zygote of something extrinsic to the species. Heredity, on the other hand, deals with the {183}transmission of something intrinsic from gamete to zygote and from zygote to gamete. It is the participation of the gamete in the process that is our criterion of what is and what is not heredity.

Better hygiene and better education, then, are good for the zygote, because they help him to make the fullest use of his inherent qualities. But the qualities themselves remain unchanged in so far as the gamete is concerned, since the gamete pays no heed to the intellectual development of the zygote in whom he happens to dwell. Nevertheless, upon the gamete depend those inherent faculties which enable the zygote to profit by his opportunities, and, unless the zygote has received them from the gamete, the advantages of education are of little worth. If we are bent upon producing a permanent betterment that shall be independent of external circumstances, if we wish the national stock to become inherently more vigorous in mind and body, more free from congenital physical defect and feeble mentality, better able to assimilate and act upon the stores of knowledge which have been accumulated through the centuries, then it is the gamete that we must consult. The saving grace is with the gamete, and with the gamete alone.

People generally look upon the human species as having two kinds of individuals, males and females, and it is for them that the sociologists and legislators frame their schemes. This, however, is but an imperfect view to {184}take of ourselves. In reality we are of four kinds, male zygotes and female zygotes, large gametes and small gametes, and heredity is the link that binds us together. If our lives were like those of the starfish or the sea-urchin, we should probably have realised this sooner. For the gametes of these animals live freely, and contract their marriages in the waters of the sea. With us it is different, because half of us must live within the other half or perish. Parasites upon the rest, levying a daily toll of nutriment upon their hosts, they are yet in some measure the arbiters of the destiny of those within whom they dwell. At the moment of union of two gametes is decided the character of another zygote, as well as the nature of the population of gametes which must make its home within him. The union once affected the inevitable sequence takes its course, and whether it be good, or whether it be evil, we, the zygotes, have no longer power to alter it. We are in the hands of the gamete; yet not entirely. For though we cannot influence their behaviour we can nevertheless control their unions if we choose to do so. By regulating their marriages, by encouraging the desirable to come together, and by keeping the undesirable apart we could go far towards ridding the world of the squalor and the misery that come through disease and weakness and vice. But before we can be prepared to act, except, perhaps, in the simplest cases, we must learn far more about them. At present we are woefully ignorant {185}of much, though we do know that full knowledge is largely a matter of time and means. One day we shall have it, and the day may be nearer than most suspect. Whether we make use of it will depend in great measure upon whether we are prepared to recognise facts, and to modify or even destroy some of the conventions which we have become accustomed to regard as the foundations of our social life. Whatever be the outcome, there can be little doubt that the future of our civilisation, perhaps even the possibility of a future at all, is wrapped up with the recognition we accord to those who live unseen and inarticulate within us—the fateful race of gametes so irrevocably bound to us by that closest of all ties, heredity.

{187}

APPENDIX

As some readers may possibly care to repeat Mendel's experiments for themselves, a few words on the methods used in crossing may not be superfluous. The flower of the pea with its standard, wings, and median keel is too familiar to need description. Like most flowers it is hermaphrodite. Both male and female organs occur on the same flower, and are covered by the keel. The anthers, ten in number, are arranged in a circle round the pistil. As soon as they are ripe they burst and shed their pollen on the style. The pollen tubes then penetrate the stigma, pass down the style, and eventually reach the ovules in the lower part of the pistil. Fertilisation occurs here. Each ovule, which is reached by a pollen tube, swells up and becomes a seed. At the same time the fused carpels enclosing the ovules enlarge to form the pod. When this, the normal mode of fertilisation, takes place, the flower is said to be **selfed**.

In crossing, it is necessary to emasculate a flower on the plant chosen to be the female parent. For this purpose a young flower must be taken in which the anthers have not yet burst. The keel is depressed, and the stamens bearing the anthers are removed at their base by a {188}pair of fine forceps. It will probably be found necessary to tear the keel slightly in order to do this. The pistil is then covered up again with the keel, and the flower is enclosed in a bag of waxed paper until the following day. The stigma is then again exposed and dusted with ripe pollen from a flower of the plant selected as the male parent. This done, the keel is replaced, and the flower again enclosed in its bag to protect it from the possible attentions of insects until it has set seed. The bag may be removed in about a week after fertilisation. It is perhaps hardly necessary to add that strict biological cleanliness must be exercised during the fertilising operations. This is readily attained by sterilising fingers and forceps with a little strong spirit before each operation, thereby ensuring the death of any foreign pollen grains which may be present.

The above method applies also to sweet peas, with these slight modifications. As the anthers ripen relatively sooner in this species, emasculation must be performed at a rather earlier stage. It is generally safe to choose a bud about three parts grown. The interval between emasculation and fertilisation must be rather longer. Two to three days is generally sufficient. Further, the sweet pea is visited by the leaf-cutter bee, *Megachile*, which, unlike the honey bee, is able to depress the keel and gather pollen. If the presence of this insect is suspected, it is desirable to guard against the risk of admixture of {189}foreign pollen by selecting for pollinating purposes a flower which has not quite opened. If the standard is not erected, it is unlikely to have been visited by *Megachile*. Lastly, it not infrequently happens that the little beetle *Meligethes* is found inside the keel. Such flowers should be rejected for crossing purposes.

{191}

www.ingramcontent.com/pod-product-compliance
Lightning Source LLC
Chambersburg PA
CBHW031927240526
45464CB00023B/1867